Cambridge astrophysics series

Gamma-ray astronomy

Gamma-ray astronomy

POOLLA V. RAMANA MURTHY FNA
Professor of Physics, Tata Institute of Fundamental Research, Bombay

ARNOLD W. WOLFENDALE FRS
Professor of Physics, University of Durham

The right of the University of Cambridge to print and sell all manner of books was granted by Henry VIII in 1534. The University has printed and published continuously since 1584.

CAMBRIDGE UNIVERSITY PRESS
Cambridge
London New York New Rochelle
Melbourne Sydney

Published by the Press Syndicate of the University of Cambridge
The Pitt Building, Trumpington Street, Cambridge CB2 1RP
32 East 57th Street, New York, NY 10022, USA
10 Stamford Road, Oakleigh, Melbourne 3166, Australia

First published 1986

Printed in Great Britain at the University Press Cambridge

British Library cataloguing in publication data
Ramana Murthy, Poolla V.
1. Gamma-ray astronomy
I. Title II. Wolfendale, A. W.
522′.686 QB471

Library of Congress cataloguing in publication data
Ramana Murthy, P. V.
Gamma-ray astronomy
(Cambridge astrophysics series ;)
Bibliography: p.
Includes index.
1. Gamma-ray astronomy. I. Wolfendale, A. W.
II. Title III. Series.
QB471.W65 1986 522′.686 85-20940

ISBN 0 521 30528 4

PN

To Subhadra and Audrey

CONTENTS

PREFACE

Gamma-ray astronomy comprises the view of the Universe through what is essentially the last of the electromagnetic windows to be opened. All other windows from radio right through to X-rays have already been opened wide, and as is well known their respective astronomies are quite well developed – and the views there are very rich. Gamma-ray astronomy promises to be likewise; the strong link of γ-rays to very energetic processes and the considerable penetration of the γ-rays see to that.

Admittedly one deals with a small number of photons in this new window and yet a considerable amount of progress has already been made; hopefully this progress will shine through in what follows.

It is usually necessary to make a selection of topics when writing a book, and the present one is no exception. The selection made here reflects both the interests of the authors (both of whom are cosmic ray physicists) and the perceived needs of the subject. The authors' interests and, no doubt, biases show through in the areas in which they have themselves contributed (Chapters 4 and 5). There appears to be a contemporary need for a comprehensive review of gamma-ray bursts and this is the reason for an extended Chapter 3. We have not included in Chapter 2 any material relating to gamma-ray lines in solar flares – a very important subject in its own right – as we felt that it was outside the character of this book, dealing as it does with source regions exclusively beyond the solar system. Furthermore, we have not reproduced many formulae, tables or graphs concerning γ-ray production mechanisms; we touch upon them only briefly. The mechanisms have a character of unchangeability associated with them; nor has anything significantly new been added in recent years. Reference is therefore made in the text to some of the books and original articles where detailed information on the subject can be found. Similarly, we have not elaborated on the detectors used in the experiments, but confined ourselves only to a cursory description of a few of them. Rather we have devoted much of the book to a presentation of experimental facts and what one makes of them at the current stage of development.

While undergraduates may find the present book useful as a supplement to earlier publications, it is the authors' view that it will be more useful to graduate students and young researchers who have spent a few years carrying out research in the area.

Hopefully, the book will also be of value to the wide fraternity of astronomers who work in the other regions of the electromagnetic spectrum. Already gamma-ray astronomy is having something of interest to say about all these regions, as well as requiring information from them, and the general astronomical value of the subject will no doubt increase over the next few years.

We have benefited considerably from discussions with many colleagues and fellow workers in the field, too numerous to mention individually, and our thanks go out to them.

Mention must be made, however, of a number of friends who contributed specifically to the production of the manuscript of the book. Mrs Pauline Russell drew the diagrams, and the typing was carried out by Mrs Margaret Chipchase, and we are grateful to them for their effort and expertise. Mrs C. V. Raisinghani and Mr H. Fernandez are also thanked for their help with word processing.

Special thanks must go to Dr Chaman L. Bhat, himself a noted worker in the γ-ray field, for his painstakingly reading, correcting and improving the manuscript. His contribution went far beyond the usual proof-reader's work and we are most grateful to him.

P. V. Ramana Murthy is also grateful to the Director of the Tata Institute of Fundamental Research, Professor B. V. Sreekantan, for his support.

1

PRODUCTION AND ABSORPTION MECHANISMS

1.1 Introduction

In this chapter attention will be given to the various production and absorption mechanisms operating in the celestial settings. Although the value of the subject is bound up with these two topics – production is by way of a variety of very energetic processes, and the low absorption experienced by γ-rays allows us to 'see' regions not otherwise accessible – they are not treated in great detail here. The reason is that the processes are rather well known and have been described in detail by a number of authors. Specifically, the books of Stecker (1971), Chupp (1976) and Hillier (1984) give excellent treatments. Our own descriptions, then, are brief.

Starting with production mechanisms, a summary is given in Figure 1.1. Understandably, the relative importance of the various mechanisms depends on the properties of the production region: gas density, temperature, magnetic field, ambient radiation etc. It is often the determination of these conditions that is the end product of the analysis of the γ-ray observations.

1.2 Gamma-ray production mechanisms

1.2.1 Gamma-ray lines

γ-ray lines have been observed from a variety of regions: solar flares, the Galactic Centre, Galactic Plane and the object SS 433, and they exhibit a variety of temporal features, from time independence to rapid time variability.

The lines generated in solar flares are proving to be of considerable

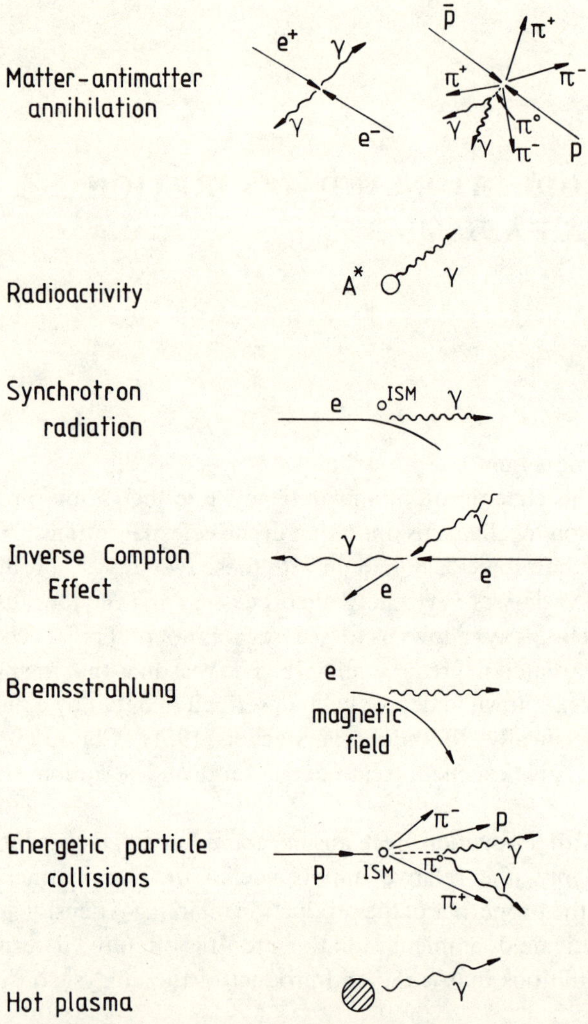

Figure 1.1. Summary of production mechanisms. A* represents an excited nucleus. The other symbols have their usual meanings.

interest, but in view of our preoccupation with astronomical regions further afield we refer the reader to Ramaty and Lingenfelter (1981), Ramaty *et al.* (1982) and references quoted therein for such discussions, and more elsewhere.

γ-ray lines are produced in (*a*) e^+e^- annihilations, (*b*) neutron capture reactions, (*c*) the de-excitation of target nuclei struck by energetic protons and other particles, together with projectile nuclei excited by impact, or by virtue of the fact that they are the radioactive products of stellar ejecta, and (*d*) Landau transitions. These topics are described in turn.

(*a*) The best known line is the annihilation line at 511 keV resulting from an e^+e^- collision at rest:

$$e^+ + e^- \rightarrow \gamma + \gamma \tag{1.1}$$

The annihilation can also take place while the particles are in flight (free annihilation), in which case the resulting γ-rays will have a continuous energy distribution instead of being a line (see Section 1.2.2 below). At low energies the e^+ and e^- first form positronium atoms, 25% of which are in the singlet $^1S_0$ state, decaying by the mode as shown in equation (1.1) with a mean lifetime of 1.25×10^{-10} s. The resulting γ-rays appear as lines at 511 keV. The remaining 75% of positronium atoms are in the triplet $^3S_1$ state and decay into three γ-rays with a lifetime of 1.5×10^{-7} s; these γ-rays form a continuum with a maximum energy of 511 keV.

The γ-ray source strength is proportional to the product of the e^+ and e^- densities. The positrons themselves arise from β-decays of nuclides produced in nucleosynthesis processes or from $\pi^+ \rightarrow \mu^+ \rightarrow e^+$ decays, the parent pions having been produced in the interactions of energetic particles with ambient matter, or through $\bar{p}p$ annihilations. Positrons may also be produced in more exotic (but not necessarily unrealistic) processes going on near massive black holes (Blandford 1976, Lovelace 1976, Lingenfelter *et al.* 1978) or around mini-primordial black holes (Okeke and Rees 1980).

(*b*) Thermalised neutrons can produce γ-rays at 2.22 MeV by the capture reaction

$$n + {}^1H \rightarrow \gamma + {}^2H \tag{1.2}$$

the cross section for which is $7.3 \times 10^{-20}\ v^{-1}\ cm^2$, where v is the velocity of the neutron in units of cm s^{-1}. The neutrons themselves originate in the interactions of energetic particles with the ambient matter, e.g. in reactions of the type ^{4}He (p, pn)^{3}He, ^{2}H(d, n)^{3}He and ^{3}H(d, n)^{4}He, and later become thermalised by repeated collisions with hydrogen. For the reaction (1.2) to take place, the density of hydrogen must be greater than 10^{16} cm^{-3}, for otherwise the neutrons would decay before interacting. It is therefore likely that this mechanism operates in relatively dense regions such as in solar flares and accretion disks but not in the interstellar medium. Neutron capture by many nuclei gives rise to compound nuclei

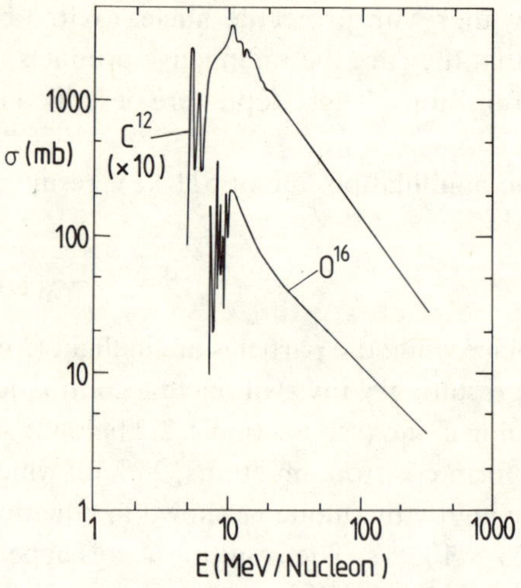

Figure 1.2. The interaction cross sections, leading to γ-ray emission, for protons on ^{12}C and ^{16}O as a function of proton kinetic energy (after Ramaty and Lingenfelter 1975).

which decay by γ-emission, and are thus of interest; particularly relevant is the capture of neutrons by ^{56}Fe leading to excited states of ^{57}Fe which emit γ-rays at 7.632 and 7.646 MeV. The surface regions of neutron stars are believed to be rich in iron and these are potentially important sites for the above process to occur.

(*c*) Turning to the interesting area of cosmic ray-induced lines, it is instructive to make an order of magnitude estimate of the expected flux of the line resulting from protons interacting with an important nuclear component of the ISM such as oxygen. The cross section for ^{16}O as a function of proton kinetic energy is given in Figure 1.2 (and also that for ^{12}C); it is seen that protons in the range 8–30 MeV are largely responsible, yielding a mean $\sigma \approx 100$ mb. The proton intensity in the ISM in this energy range is virtually unknown but we take 2 cm^{-2} s^{-1} sr^{-1} as illustrative. The cosmic abundance of ^{16}O is $\approx 7 \times 10^{-4}$ of hydrogen by number and the corresponding number of excited oxygen atoms produced for a region of the ISM, where the total gas density is 1 cm^{-3}, follows as $\approx 2 \times 10^{-27}$ cm^{-3} s^{-1}. If each excited oxygen atom were to de-excite by way of a γ-ray line then the line flux expected along a path length of 10 kpc would be 5×10^{-6} cm^{-2} s^{-1} sr^{-1}. This is clearly a very low flux, giving a counting rate of only ≈ 1 γ-ray per day in a detector of area 100 cm^2 and solid angle (1/60) sr (the γ-ray flux will be strongly collimated in the Galactic Plane). Detailed calculations in the above manner, but with much greater accuracy, have been made by Meneguzzi and Reeves (1975) and Ramaty *et al.* (1979). The most conspicuous line expected appears to be the one at 6.129 MeV resulting from ^{16}O, but there are many others. The results of the calculations by Ramaty *et al.* are given later in Figure 2.1.

An important feature is the line width and this can be examined briefly. There will be Doppler broadening depending on the lifetime of the excited state and on the density and temperature of the medium, due to the kinetic energy imparted to the nuclide in the collision and the general Maxwell–Boltzmann distribution of the velocities of the nuclide. The line width from recoil often dominates, as can be seen simply from the considerations of the previous section. A 10 MeV proton has $\beta = 0.14$ and a recoil oxygen nucleus will have $\beta \approx 0.01$. If the oxygen nucleus is free in the ISM it will decay before being reduced to rest and the line width will be of order 1 %, i.e. 60 keV for a line at 6 MeV. An even larger line width will arise from those cosmic ray nuclei (carbon, oxygen etc.) which are excited by collision with ISM nuclei.

Much of the 'heavy material' (C, O etc.) in the ISM is in the form of grains, however, and here the struck nuclei will usually come to rest before decaying; the line widths then will be very narrow, less than the instrumental resolution ($\approx$5 keV for a good detector). (Figure 2.2 shows these features.)

In addition to the ^{16}O line at 6.129 MeV, there are other important ones, notably 4.438 MeV from ^{12}C; 2.313 and 5.105 MeV from ^{14}N; 2.741, 6.917 and 7.117 MeV from ^{16}O; 1.634, 2.613 and 3.34 MeV from ^{20}Ne; 1.369 and 2.754 MeV from ^{24}Mg; 1.779 and 6.878 MeV from ^{28}Si; and 0.847, 1.238 and 1.811 MeV from ^{56}Fe. Long lived nuclides from supernova ejecta will also contribute (Clayton *et al.* 1969, Ramaty and Lingenfelter 1981). Examples are the lines at 1.809 MeV from ^{26}Al; 0.847 and 1.238 MeV from ^{56}Co; 1.156, 0.076 and 0.068 MeV from ^{44}Ti; and 1.332 and 1.173 MeV from ^{60}Co.

(*d*) In regions where strong magnetic fields exist, cyclotron emission lines can be produced from transitions between Landau levels. Transitions from the first such level result in the emission of a γ-ray line whose energy is given by

$$E_\gamma(\text{keV}) = heB/m_e c = 11.6\, B_{12} \tag{1.3}$$

where B_{12} is the magnetic field in units of 10^{12} G. The line may be seen in absorption or emission depending on the temperature of the ambient medium.

Before closing this discussion, we should examine a number of other possible causes of the displacement or energy smearing of γ-ray lines.

Although the decay of a neutral pion at rest results in two γ-rays of half the rest mass (i.e. 67.48 MeV) each, the π^0 are seldom produced at rest in the interactions of energetic particles with matter because of motion of the centre-of-mass in the laboratory frame. Consequently, the γ-ray 'line' has a broad hump around 67.5 MeV. Gravitational red shift can also be very large: for example, a γ-ray of energy E_γ produced close to a neutron star can be gravitationally red-shifted by as much as $\Delta E_\gamma = (GM/R)(E_\gamma/c^2) \approx (0.1 \text{ to } 0.3)E_\gamma$; see the discussion in Sections 3.3.1 and 3.4.2B with reference to the Jacobson transient and the γ-ray burst of November 19, 1978, respectively. Furthermore, the cosmological red-shifts operating on γ-rays produced in the remote past and in distant regions of the Universe but reaching us now, may reduce the energy of the γ-ray considerably; see Stecker (1971, 1973). Finally, Doppler shifts due to the organised motion of the bulk of emitting nuclides can, of course, make the lines appear at lower *or* at higher energies than those at emission. Experimental evidence for such Doppler shifts of the 1.369 MeV line identified with ^{24}Mg from SS 433 has been presented recently by Lamb *et al.* (1983); see Section 2.4.2.

1.2.2 The gamma-ray continuum

1.2.2A Electrons

The particles responsible for the continuum are, principally, cosmic ray electrons and protons.

Starting with electrons, continuum γ-rays are produced by (*a*) annihilation in flight with positrons, (*b*) bremsstrahlung in the Coulomb fields of nuclei and electrons, (*c*) inverse Compton scattering against the ambient radiation, (*d*) synchrotron emission in magnetic fields, and (*e*) curvature radiation in the magnetospheres of pulsars.

(a) *Annihilation in flight*. The cross section for e^+e^- annihilation at extreme relativistic energies is given by:

$$\sigma_A = \frac{\pi r_e^2}{\gamma}[\ln(2\gamma) - 1] \tag{1.4}$$

(see Stecker 1971 for details), where $\gamma\ (=E_e/m_e c^2)$ is the Lorentz factor of the positrons and $r_e\ (=e^2/m_e c^2 = 2.82 \times 10^{-13}$ cm) is the classical radius of the electron. Two-photon emission is the dominant result of annihilation, leading to a continuum of photon energies. At high positron energies, the photon emitted in the extreme forward direction has almost all the energy of the positron while the other photon has an energy of $m_e c^2/2 = 0.256$ MeV.

The cross section given by equation (1.4) is quite low; as an example, for a 100 MeV positron, $\sigma_A = 6 \times 10^{-27}$ cm^2, leading to a mean free path for annihilation of $1.6 \times 10^{26}/n_e$ cm, where n_e is the electron number density. Now, in the ISM, n_e is $\approx 3 \times 10^{-2}$ cm^{-3} and the mean free path follows as $\approx 5 \times 10^{27}$ cm, i.e. nearly 2000 Mpc; in turn, the mean life of e^+ in the Galaxy would be 5×10^9 y. Other processes will clearly be more important in degrading the positron energy, although eventually, for those positrons which do not escape from the Galaxy but rather slow down to very low energy, the much higher annihilation cross section at these low energies will cause the positrons to disappear.

(b) *Bremsstrahlung*. The differential cross section for an electron of energy E_e to radiate a photon with energy between E_γ and $E_\gamma + dE_\gamma$ in the Coulomb field of a nucleus with charge Z is given by (Rossi 1952)

$$\sigma(E_e, E_\gamma)\, dE_\gamma = 4\alpha Z^2 r_e^2 \frac{dE_\gamma}{E_\gamma} F(E_e, \nu) \tag{1.5}$$

where

$$F(E_e, \nu) = \left[1 + (1-\nu)^2 - \frac{2}{3}(1-\nu)\right]\left[\ln\left(\frac{2E_e}{m_e c^2}\frac{1-\nu}{\nu}\right) - \frac{1}{2}\right] \tag{1.6}$$

for the case of no screening (i.e. a bare nucleus); i.e. $E_e \ll 100 m_e c^2[\nu/(1-\nu)]Z^{-1/3}$, and

$$F(E_e, \nu) = [1 + (1-\nu)^2 - \tfrac{2}{3}(1-\nu)] \ln(183 Z^{-1/3}) + \tfrac{1}{9}(1-\nu) \tag{1.7}$$

for the case of complete screening; i.e. $E_e \gg 100 m_e c^2(\nu/1-\nu)Z^{-1/3}$. Here α is the fine structure constant and $\nu(=E_\gamma/E_e)$ is the fractional energy carried by the photon. The bremsstrahlung photons have the spectral shape $\sim 1/E_\gamma$, so that more photons are emitted at lower energies. For ionised hydrogen, obviously the case of no screening (equation (1.6)) applies. Bremsstrahlung in the Coulomb field of the atomic electrons is included by replacing Z^2 by $Z(Z+1)$ in equation (1.5). Bremsstrahlung

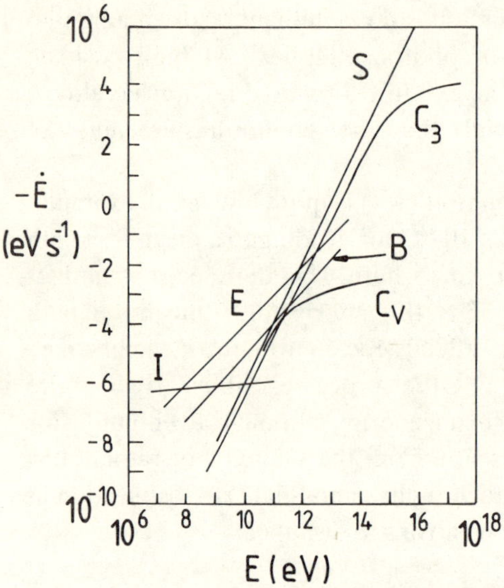

Figure 1.3. Rate of energy loss of electrons $-\dot{E}$ as a function of electron energy E (from the work of Ramaty 1974). Many of the processes involve the production of γ-rays (or X-rays when E is 'small'). S: synchrotron losses in a magnetic field of 4 μG; C_v: inverse Compton interactions with visible photons (energy density = 0.45 eV cm^{-3}); C_3: ditto with '3 K' microwave background; B: bremsstrahlung (gas density = 1 atom cm^{-3}); E: escape, with mean life in galaxy of 3.3 $\times$ 10^6 y; I: ionisation loss. (E and I do not involve γ-ray production.)

in the field of the electron assumes importance when the numbers of $e^{\pm}$ are far higher than those of the nuclei as in regions where temperatures are in excess of 10^{10} K.

The mechanism of radiation loss is such that the electron falls in energy by a factor e in one radiation length; the radiation length in hydrogen is $\approx$61 g cm^{-2} and for the ISM, with a mean density 1 atom cm^{-3}, this length is about 10 Mpc. In turn, for an electron moving with $v \approx c$,the mean life for loss by bremsstrahlung is $\approx$40 My. Figure 1.3 summarises the situation for loss of electron energy by bremsstrahlung and also some of the other processes to be described.

Stecker (1975) has shown that in the case of bremsstrahlung from electrons in the interstellar gas, the γ-ray emissivity, $q_b(E_\gamma)$, is given by

$$q_b(E_\gamma) = 4.3 \times 10^{-25} n I_e(>E_\gamma)/E_\gamma \text{ cm}^{-3}\text{ s}^{-1}\text{ MeV}^{-1} \tag{1.8}$$

where n is the number density of nuclei in the production region and

$I_e(>E_\gamma)$ is the integral energy spectrum of the electrons. Equation (1.8) shows that the exponent of the bremsstrahlung γ-ray spectrum has the same value as that of the parent electron spectrum.

An order of magnitude calculation can be made for the typical path length of ~10 kpc and $n = 1$ atom cm^{-3} encountered in the Galactic Plane. Strong and Wolfendale (1981) have shown that $I_e(>100\text{ MeV})$ is $\approx 5 \times 10^{-3}\text{ cm}^{-2}\text{ s}^{-1}\text{ sr}^{-1}$, yielding $q_b(100\text{ MeV}) = 2 \times 10^{-27}\text{ cm}^{-3}\text{ s}^{-1}\text{ MeV}^{-1}$ and $q_b(>100\text{ MeV}) \approx 2 \times 10^{-26}\text{ cm}^{-3}\text{ s}^{-1}\text{ sr}^{-1}$. In fact, as will be demonstrated later, a similar contribution comes from proton interactions. We see that the flux is about 100 times that expected for a (strong) γ-ray line (see Section 1.2.1) and, when it is appreciated that γ-ray detectors in the 100 MeV region can have a typical area of 500 cm^2, solid angle of 0.25 sr and detection efficiency of $\approx$10%, the expected counting rate will now be 'reasonable': $\approx$500 γ-rays per day (multiplied by two to allow for proton-induced γ-rays).

(c) Inverse Compton scattering. An ambient photon of energy ε appears to be moving with an energy $\gamma\varepsilon$ in the rest system of a relativistic electron with Lorentz factor γ. In this inertial frame, the Compton scattered photon has an energy $\lesssim \gamma\varepsilon$ and, when transformed back to the laboratory system, an energy $\sim\gamma^2\varepsilon$. The energy of the Compton boosted photon is given by

$$E_\gamma \approx \varepsilon\gamma^2 \quad \text{when} \quad \gamma\varepsilon \ll m_e c^2 \tag{1.9}$$

and

$$E_\gamma \sim E_e \quad \text{when} \quad \gamma\varepsilon \gg m_e c^2 \tag{1.10}$$

This process is important when the density of ambient photons is high. It is also an efficient process in elevating photon energies to very high levels. Cross sections for the regions represented by equations (1.9) and (1.10) are given, respectively (Heitler 1960), by:

$$\sigma_c = \sigma_T\left(1 - \frac{2\gamma\varepsilon}{m_e c^2}\right) \tag{1.11}$$

and

$$\sigma_c = \frac{3}{8}\,\sigma_T\left(\frac{m_e c^2}{\gamma\varepsilon}\right)\left[\ln\left(\frac{2\gamma\varepsilon}{m_e c^2}\right) + \frac{1}{2}\right] \tag{1.12}$$

Here $\sigma_T(=8\pi r_e^2/3)$ is the Thomson cross section, having a numerical value of 6.65×10^{-25} cm^2.

The condition $\gamma\varepsilon \ll m_e c^2$ is commonly experienced in connection with the interaction of electrons with the 2.7 K background radiation. Here the mean value of ε, $\langle\varepsilon\rangle$, is $\approx 6 \times 10^{-4}$ eV and the condition is therefore satisfied for $\gamma < 10^9$, i.e. electron energies below about 5×10^{14} eV.

For such energies, the accurate relation between the mean γ-ray energy, $\langle E_\gamma\rangle$, and $\langle\varepsilon\rangle$ is

$$\langle E_\gamma\rangle = \tfrac{4}{3}\gamma^2\langle\varepsilon\rangle \tag{1.13}$$

Assuming that the differential energy spectrum of electrons is given by

$$I_e(E_e) = KE_e^{-\Gamma} \tag{1.14}$$

where K is a normalisation factor, Stecker (1975) has shown that the γ-ray emissivity, $q_c(E_\gamma)$, is given by

$$q_c(E_\gamma) = \tfrac{8}{3}\pi\sigma_T\rho_{ph}(m_e c^2)^{1-\Gamma}(\tfrac{4}{3}\langle\varepsilon\rangle)^{[(\Gamma-3)/2]}KE_\gamma^{-[(\Gamma+1)/2]} \tag{1.15}$$

Here ρ_{ph} is the energy density of the low energy photons (0.24 eV cm^{-3} for the microwave background) and all other quantities are as defined earlier. The exponent of the γ-ray energy spectrum, given by $-(\Gamma + 1)/2$, is flatter than that of the parent electron spectrum for values of $\Gamma > 1$.

Insofar as the cosmic ray electron spectrum in the ISM has a power law spectrum satisfying this condition, the inverse Compton mechanism can be an important source of low energy γ-rays. For example, 1 MeV γ-rays will come from electrons of $\approx 2 \times 10^{10}$ eV (equation (1.13)) and here $\Gamma \approx 3$.

Figure 1.3 includes the role of inverse Compton interactions in dissipating electron energy.

(d) *Synchrotron emission*. Photons are emitted when an electron (or positron) traverses a transverse magnetic field (see Ginzburg and Syrovatskii (1964) for an exhaustive treatment of the subject). In a homogeneous magnetic field of strength H an electron of energy E_e moving at an angle θ with respect to the field will emit synchrotron radiation of intensity (energy flux density per unit frequency interval) along the direction of observation given by

$$I_\nu = \frac{3^{1/2}e^3}{m_e c^2} N_e(\mathbf{K})H \sin\theta \, \frac{\nu}{\nu_c} \int_{\nu/\nu_c}^{\infty} K_{5/3}(\eta)\, d\eta \tag{1.16}$$

Here e is the electron charge; $N_e(\mathbf{K})$ is the number of electrons per unit solid angle along the line of vision whose velocity vectors are towards the observer; $K_{5/3}$ is the modified Bessel function of the second kind; ν is the frequency of the emitted photon and

$$\nu_c = \frac{3eH \sin\theta}{4\pi m_e c}\left(\frac{E_e}{m_e c^2}\right)^2 \tag{1.17}$$

The maximum emission occurs at a frequency given by

$$\nu_m = 1.2 \times 10^6 H_\perp \left(\frac{E_e}{m_e c^2}\right)^2 \tag{1.18}$$

or, equivalently, at an energy

$$E_{\gamma,m}(\text{eV}) = h\nu_m = 5 \times 10^{-9} H_\perp \left(\frac{E_e}{m_e c^2}\right)^2 \tag{1.19}$$

Here $H_\perp = H \sin\theta$ is in gauss. The energy of the electron is halved in a time $t_{1/2}$ given by

$$t_{1/2} = 5 \times 10^8 \left(\frac{m_e c^2}{E_e}\right)\left(\frac{1}{H_\perp(\text{G})}\right)^2 \text{ s} \tag{1.20}$$

and the corresponding rate of energy loss, for a particular magnetic field strength, is shown in Figure 1.3.

If the radiating electrons have a differential energy spectrum given by equation (1.14) the synchrotron photon *number* spectrum will be of the form

$$G_s(E_\gamma) \propto E_\gamma^{-[(\Gamma+1)/2]} \tag{1.21}$$

a form similar to that of the inverse Compton produced photons.

Attention must be drawn here to the fact that in other windows of the electromagnetic radiation spectra are usually expressed in terms of the power radiated. To find the exponent of the radiated power spectrum (W m^{-2} Hz^{-1}), one has to multiply the number spectrum equation (1.21) by E_γ, leading to the resulting synchrotron power spectrum:

$$J_s(E_\gamma) \propto E_\gamma^{-[(\Gamma-1)/2[} \tag{1.22}$$

(e) *Curvature radiation*. In the exceptionally strong magnetosphere of a pulsar, high energy electrons and positrons are constrained to move parallel to the magnetic field lines with an essentially zero pitch angle. Since the field lines themselves are curved, the particles radiate 'curvature' radiation in their direction of motion (Manchester and Taylor 1977). The characteristic energy of this radiation, E_c, is given by

$$E_c(\text{eV}) \approx \frac{3}{2}\frac{\hbar c\gamma^3}{\rho_c} = \frac{2.96 \times 10^{-5}\gamma^3}{\rho_c(\text{cm})} \tag{1.23}$$

Here ρ_c is the radius of curvature of the magnetic field line, $\hbar = h/2\pi$ and $\gamma = E_e/m_e c^2$. This process is important for very-high energy electrons and positrons in the environs of pulsars. As an example, a 10^{13} eV electron moving along a field line with a curvature of 10^8 cm, typical for a pulsar, emits photons of energy $\approx$2.5 GeV.

1.2.2B Protons

High energy protons can, in principle, produce γ-rays by processes (*b*) to (*e*) listed in 1.2.2A for electrons. These mechanisms are, however, very inefficient because of the much higher mass-to-charge ratio for the proton ($m_p/m_e = 1836$). Protons can, on the other hand, generate π^0 mesons in inelastic collisions with matter either directly or indirectly, and less efficiently, through the decays of kaons and hyperons and through the annihilations of antinucleons (kaons, hyperons and antinucleons being produced in the inelastic collisions). Protons can also interact with ambient radiation, e.g. the microwave background to produce π^0 mesons. These neutral pions decay into γ-rays with a mean lifetime of γ_π 0.83×10^{-16} s, where γ_π is the Lorentz factor of the pion.

Interaction with matter. The threshold kinetic energy, T_{th}, for producing a particle with a mass m in a proton–proton collision is given by

$$T_{th} = 2mc^2(1 + m/4m_p) \tag{1.24}$$

where m_p is the proton mass. For a π^0 meson of mass 134.96 MeV/c^2, the threshold kinetic energy is 279.6 MeV. If the target is a heavy nucleus instead of a proton the threshold is lowered somewhat due to the Fermi motion of nucleons within the nucleus. The product of the production cross section, σ_{π^0}, and the neutral pion multiplicity, ζ_{π^0}, is shown as a function of proton energy in Figure 1.4. Folding this dependence on energy with the differential proton energy spectrum (Fig. 1.5), it is found that the bulk of the neutral pions are produced by protons having kinetic energies between 1 and 10 GeV. Quantitatively, the situation is as shown in Figure 1.6, taken from extensive calculations by Stephens and Badhwar (1981). We note that only 10 % of all pions are produced by protons of energy above 10 GeV. Also shown in Figure 1.6 are the percentages of pions produced with a low energy, namely below 0.25 and 0.5 GeV; here a significant fraction comes from protons of kinetic energy below 1 GeV,

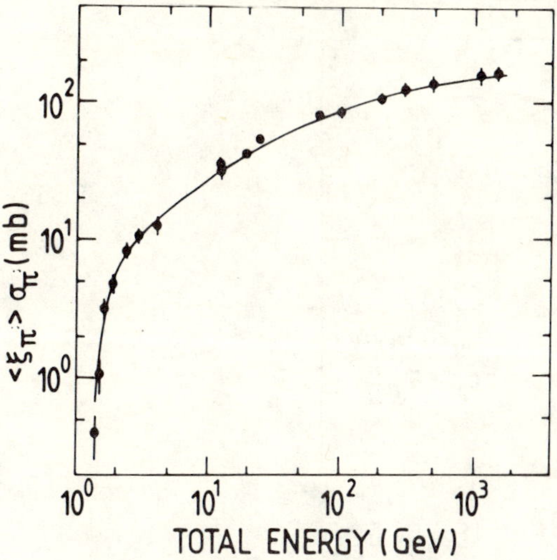

Figure 1.4. Product of cross section and multiplicity for neutral pion production in proton–proton collisions as a function of incident energy (after Stephens and Badhwar 1981).

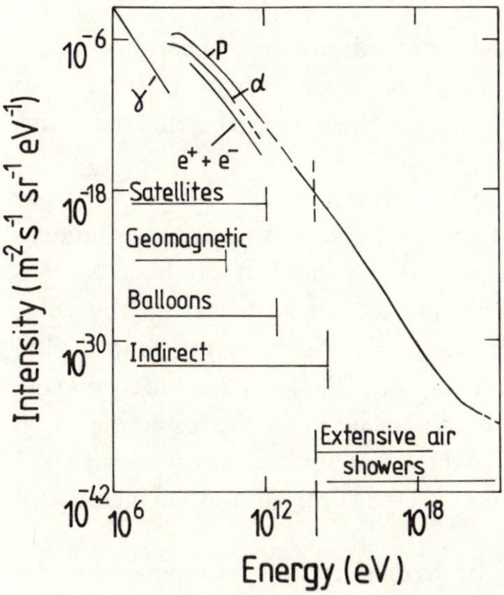

Figure 1.5. Energy spectrum of cosmic rays. In addition to the components shown there are small fluxes of heavier nuclei. The mass composition above about 10^{14} eV becomes increasingly uncertain with a possible increase in the fraction of heavy nuclei; however, there is a general agreement that the particles are mainly protons above 10^{17} eV.

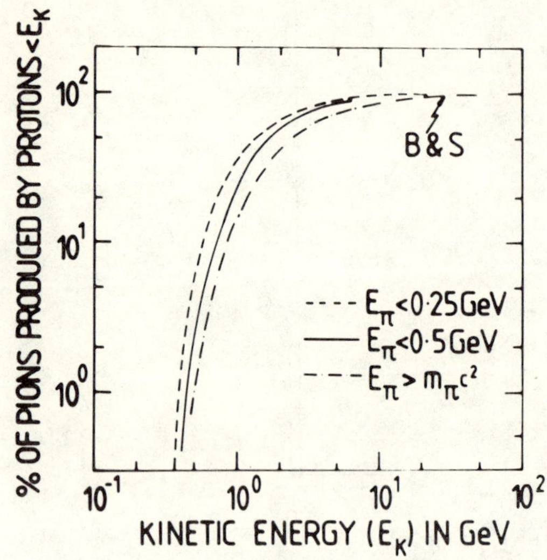

Figure 1.6. Percentage of neutral pions produced by protons of kinetic energy less than E_K as a function of E_K for various ranges of pion energy (E_π is the total pion energy). The graph is from Stephens and Badhwar (1981) (B & S).

a region of the spectrum where solar modulation causes uncertainty in the intensity estimates. Fortunately, however, most γ-ray measurements to be discussed here (Chapter 4) are of too high energy to be affected in this way.

Many workers have calculated the energy spectra of photons per hydrogen atom in the ISM, and Figure 1.7 gives results from a summary by Stephens and Badhwar. We note the variety of predictions. It is the curve for p–p (i.e. cosmic ray protons on hydrogen in the ISM) that is derived from the total inclusive cross section for neutral pion production in Figure 1.4, the kinematics of $\pi^0 \rightarrow \gamma + \gamma$ and the cosmic ray energy spectrum. To this contribution must be added the results of proton collisions with nuclei in the ISM having $Z > 1$ and cosmic rays themselves with $Z > 1$ interacting with the ISM. All the contributions are included in the upper curves.

The division of γ-ray emissivity between the various components of the cosmic ray spectrum and the ISM can be seen from the complementary work of Dodds *et al.* (1976), shown in Table 1.1, which is essentially the integral of Figure 1.7 over all γ-ray energies (and above 100 MeV) but broken down into the constituents. Case A corresponds to a primary spectrum very similar to that of M_u in Figure 1.7.

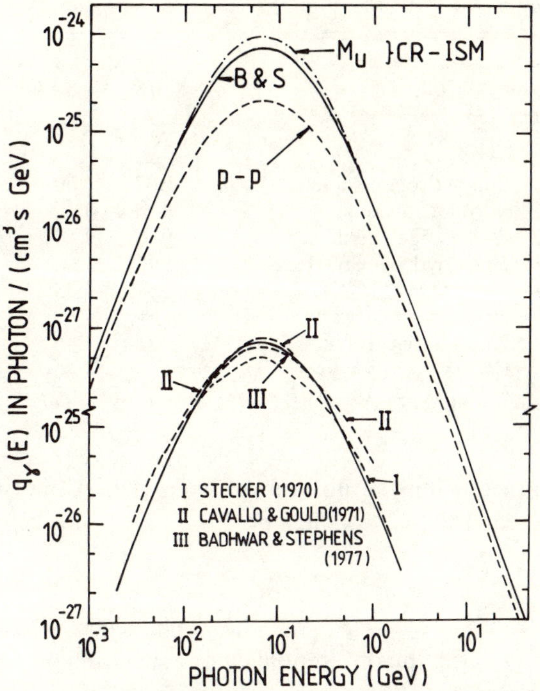

Figure 1.7. Differential production rate of γ-rays as a function of photon energy for a variety of possible cosmic ray spectra (after Stephens and Badhwar 1981). p–p relates to predictions for protons on interstellar hydrogen (using the spectrum). The remaining curves refer to all cosmic rays on all nuclei in the ISM (assuming $n_H = 1$ hydrogen atom cm^{-3}). M_u: upper spectrum adopted by SB; B & S: Badhwar and Stephens (1977). The full line in the lower set of spectra is the B & S spectrum. It is seen that there are significant differences in the predictions. A production rate intermediate between B & S and M_u is probably the best estimate at the present time.

Adopting the emissivity given in Table 1.1 for $E_\gamma > 100$ MeV we can make the standard calculation of the γ-ray intensity I_γ expected along a 10 kpc path through an ISM of 1 atom cm^{-3}. The result is $I_\gamma = 2.65 \times 10^{-25} \times 1/4\pi \times 3 \times 10^{22} = 5 \times 10^{-4}$ cm^{-2} s^{-1} sr^{-1}. This result is, quite fortuitously, similar to the electron contribution for the same γ-ray energy threshold.

Moving to much higher γ-ray energies, the predicted spectral shape is much simpler and it can be represented by a power law of roughly

Table 1.1. *Gamma-ray emissivities (after Dodds* et al. *1976)*

Photon energy	Contribution	Emissivity × $10^{-25}s^{-1}$	
		A	B
All γs	proton–hydrogen nucleus	2.17	1.60
	α–hydrogen nucleus	0.41	0.30
	$Z>2$–hydrogen nucleus	0.22	0.16
	all CR on hydrogen nuclei	2.80	2.06
	all CR on ISM	3.90	2.88
γs above 100 MeV	0.68 × emissivity for all CR on ISM	2.65	1.96

The interstellar spectra adopted are A = Fisk (1975); B = Bedijn *et al.* (1973).

constant exponent. Characterising the differential energy spectrum of cosmic rays at energies greater than a few GeV by

$$N(E_{\mathrm{p}}) \propto E_{\mathrm{p}}^{-\Gamma} \tag{1.26}$$

the multiplicity dependence as $\zeta_{\pi^0} \propto E_{\mathrm{p}}^{a}$ and the dependence of the average π^0 energy as $\langle E_{\pi^0}\rangle \propto E_{\mathrm{p}}^{b}$, the differential energy spectrum of the produced γ-rays has the form

$$q(E_\gamma) \propto E_\gamma^{-[(\Gamma+b-a-1)/b]} \tag{1.27}$$

(Stecker 1973).

If one takes the values of $\Gamma = 2.5$ and $a = \frac{1}{4}$, $b = \frac{3}{4}$ (Fermi model), the exponent of the γ-ray spectrum has a value -2.67. Although the Fermi model is not exactly applicable, as demonstrated by the accelerator experiments, the point is made that the exponent of the γ-ray energy spectrum is close to that of the parent cosmic ray spectrum.

Generation of ultra-high energy gamma-rays from proton interactions with radiation. An ambient low energy photon of energy ε, when seen in the rest system of a proton of energy $\gamma m_{\mathrm{p}}c^2$, has an energy $\gamma\varepsilon$. If $\gamma\varepsilon$ exceeds the photomeson production threshold, it can lead to π^0 production and energy degradation of the primary proton, typically by about 10 %. The threshold energy for this process is given by

$$m_{\pi^0}(1 + m_{\pi^0}/2m_{\mathrm{p}}) = 144.7\ \mathrm{MeV} \tag{1.28}$$

The 2.7 K microwave background, with photon number density of 400 cm^{-3}, forms an ambient photon field for this process to occur. The average energy of a 2.7 K photon is 6.4×10^{-4} eV, making the threshold γ as high as 2.3×10^{11}, i.e. the mechanism operates only at the extremely high proton energies of $\approx 10^{20}$ eV. Nevertheless, in the vast intergalactic distances, this process leads to energy degradation and hence to a steepening of the primary cosmic ray proton spectrum at $E_p \gtrsim 10^{20}$ eV. (This is an area of active experimental investigation in cosmic ray research.) The produced photons have energies $\approx 10^{19}$ eV, and interact with the same microwave background photons to produce e^+e^- pairs. These particles in turn produce high energy γ-rays by inverse Compton scattering. The resulting electromagnetic cascade gives γ-rays of much lower energies; see Wdowczyk *et al.* (1972) and Stecker (1973) for details.

1.2.2C $\bar{p}p$ annihilations

Neutral pions can be produced in the annihilations of protons with antiprotons. In principle, the antiprotons can arise from the interactions of energetic protons with matter, but the mechanism is too inefficient compared to the direct production of π^0 mesons during the interactions. If antimatter exists in its own right in significant quantities in the Universe, however, this process can add to others in producing neutral pions which subsequently decay into γ-rays. Stecker (1971) has calculated the differential γ-ray energy spectrum resulting from $\bar{p}p$ annihilations at rest; see Figure 1.8.

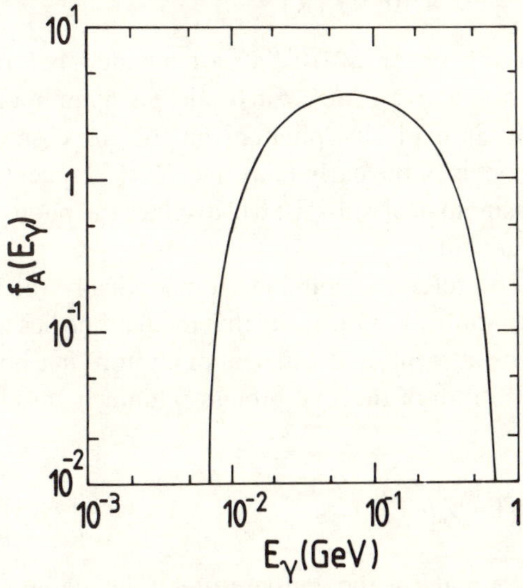

Figure 1.8. The normalised differential γ-ray spectrum from $\bar{p}p$ annihilation at rest (Stecker 1971).

1.2.2D Black body radiation and thermal bremsstrahlung

Black body radiation is the radiation emitted by a body having a large optical depth at all frequencies. Its spectrum is given by the well known Planck formula

$$I_{BB}(\nu)\,d\nu = \frac{2\pi h\nu^3}{c^2}\left[\frac{1}{\exp(h\nu/kT) - 1}\right] d\nu \text{ erg cm}^{-2}\,\text{s}^{-1} \qquad (1.29)$$

where the symbols have their usual meanings. The photon number spectrum (number of photons emitted per unit area of the source per unit time and energy interval) follows as:

$$N(E_\gamma) = 9.89 \times 10^{40} E_\gamma^2\left[\frac{1}{\exp(1.16 \times 10^{10} E_\gamma/T) - 1}\right]$$

$$\text{photons cm}^{-2}\,\text{s}^{-1}\,\text{MeV}^{-1} \qquad (1.30)$$

where T is the temperature in Kelvin and E_γ is the energy of the γ-ray in MeV.

Characteristic black body energies are: the energy corresponding to the maximum in the wavelength distribution (i.e. Wien's law)

$$E_w\,(\text{MeV}) = 4.7 \times 10^{-10}\,T\,(\text{K})$$

and the average photon energy

$$\langle E \rangle\,(\text{MeV}) = 2.7kT = 2.3 \times 10^{-10}\,T\,(\text{K})$$

Clearly, temperatures of the order of 10^{10} K are needed before appreciable production of γ-rays above 1 MeV can occur. Such temperatures are rare and occur only in explosive phenomena (on the cosmic scale) and in hot Big Bang models of the early Universe. With respect to the latter, we note that the cosmological red-shift will displace the photon energies to considerably lower values.

Equations (1.29) and (1.30) refer to emission from optically thick media (i.e. the black body situation). For optically thin media, one has to multiply equation (1.30) by the absorption coefficient of photons in a hot plasma, and the resulting spectrum of thermal bremsstrahlung γ-rays is given by

$$N(E_\gamma) \propto \frac{n_e^2}{T^{1/2}}\frac{1}{E_\gamma}\exp(-E_\gamma/kT) \qquad (1.31)$$

where n_e is the electron density and T is the temperature of the medium; for details see Hayakawa (1969) and Ginzburg (1969). Several examples of photon spectra from optically thin sources are presented in Section 3.4.1 dealing with the γ-ray bursts.

1.2.3 Summary of production mechanisms

A variety of γ-ray production mechanisms have been mentioned on the preceding pages. In any given astrophysical setting one or a few of them are more important than the others. To help appreciate the relative significances, we have listed in Table 1.2 the parameters relevant to the production of γ-rays of 1 MeV, 1 GeV and 1 TeV by the various processes.

1.3 Gamma-ray loss mechanisms

1.3.1 Introduction

It is the low loss of energy by γ-rays in penetrating a variety of astronomical environs that makes γ-rays such a powerful probe of these regions – and (the other side of the coin) that makes their detection so difficult. Hillier (1984) has given a good description of the various types of detectors and of the physical processes underlining their operation, together with the attendant problems of background γ-rays produced by unwanted processes, and the reader is referred to that book for details. Here we concentrate instead on the loss mechanisms which are operative in the astronomical environment. The topic can be divided into two parts: γ–matter interactions and γ–γ processes.

1.3.2 Gamma-ray–matter interactions

General. The relative importance of the various γ–matter interactions can be seen by reference to Figure 1.9. If we define γ-rays as photons above 0.5 MeV then only the Compton and pair production processes are important; although when we go to somewhat lower energies to provide an overlap with hard X-rays (an area of particular interest in connection with γ-ray transients), the photoelectric effect has importance, at least for modest values of Z. Nevertheless we will not describe any features of the photoelectric process here.

Compton effect. The well known theory of the Compton effect gives for the wavelength shift of the scattered photon the magnitude

$$\lambda' - \lambda = \frac{h}{m_e c}(1 - \cos\theta) \tag{1.32}$$

where λ' is the wavelength of the scattered photon, λ is the incident

Table 1.2. *Gamma-ray production parameters*

Mechanism	$E\gamma = 1\,\mathrm{MeV}$	$E\gamma = 1\,\mathrm{GeV}$	$E\gamma = 1\,\mathrm{TeV}$
Inverse Compton effect, formulae (1.13) and (1.10)	Against microwave background		
	$\varepsilon \sim 7 \times 10^{-4}\,\mathrm{eV}$	$\varepsilon \sim 7 \times 10^{-4}\,\mathrm{eV}$	$\varepsilon \sim 7 \times 10^{-4}\,\mathrm{eV}$
	$E_e \sim 1.7 \times 10^{10}\,\mathrm{eV}$	$E_e \sim 5.3 \times 10^{11}\,\mathrm{eV}$	$E_e \sim 1.7 \times 10^{13}\,\mathrm{eV}$
	Against starlight		
	$\varepsilon \sim 1\,\mathrm{eV}$	$\varepsilon \sim 1\,\mathrm{eV}$	$\varepsilon \sim 1\,\mathrm{eV}$
	$E_e \sim 4.4 \times 10^{8}\,\mathrm{eV}$	$E_e \sim 1.4 \times 10^{10}\,\mathrm{eV}$	$E_e \gtrsim 1\,\mathrm{TeV}$
	Against X-rays		
	$\varepsilon = 10\,\mathrm{keV}$	$\varepsilon = 10\,\mathrm{keV}$	$\varepsilon = 10\,\mathrm{keV}$
	$E_e \sim 4.4 \times 10^{6}\,\mathrm{eV}$	$E_e \gtrsim 1\,\mathrm{GeV}$	$E_e \gtrsim 1\,\mathrm{TeV}$
Synchrotron radiation, formula (1.19)	$B = 10^{-4}\,\mathrm{G}$		
	$E_e = 7.7 \times 10^{14}\,\mathrm{eV}$		
	$B = 1\,\mathrm{G}$	$B = 1\,\mathrm{G}$	$B = 1\,\mathrm{G}$
	$E_e = 7.2 \times 10^{12}\,\mathrm{eV}$	$E_e = 2.3 \times 10^{14}\,\mathrm{eV}$	$E_e = 7.2 \times 10^{15}\,\mathrm{eV}$
	$B = 10^{4}\,\mathrm{G}$	$B = 10^{4}\,\mathrm{G}$	$B = 10^{4}\,\mathrm{G}$
	$E_e = 7.2 \times 10^{10}\,\mathrm{eV}$	$E_e = 2.3 \times 10^{12}\,\mathrm{eV}$	$E_e = 7.2 \times 10^{13}\,\mathrm{eV}$
Bremsstrahlung	$E_e \gtrsim 2\,\mathrm{MeV}$	$E_e \gtrsim 2\,\mathrm{GeV}$	$E_e \gtrsim 2\,\mathrm{TeV}$
Curvature radiation, formula (1.23)	$\rho = 10^{8}\,\mathrm{cm}$	$\rho = 10^{8}\,\mathrm{cm}$	$\rho = 10^{8}\,\mathrm{cm}$
	$E_e = 7.7 \times 10^{11}\,\mathrm{eV}$	$E_e = 7.7 \times 10^{12}\,\mathrm{eV}$	$E_e = 7.7 \times 10^{13}\,\mathrm{eV}$
p–p inelastic collisions $\rightarrow \pi^0 \rightarrow 2\gamma$		$E_p \gtrsim 10^{10}\,\mathrm{eV}$	$E_p \gtrsim 10^{13}\,\mathrm{eV}$

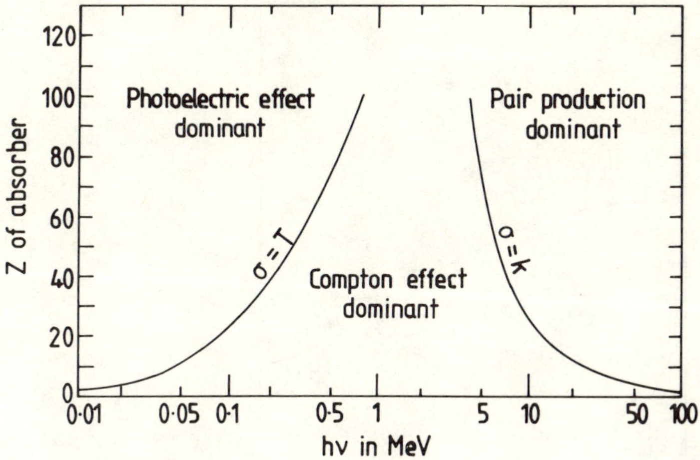

Figure 1.9. The relative importance of the major forms of γ–matter interaction as a function of γ-ray energy and atomic number of the material involved (after Evans 1955).

wavelength, θ is the angle of scatter and the other symbols have their usual meanings.

The cross section for Compton scattering depends on the polarisation of the photon beam as well as the other parameters. Following Heitler (1960) it can be written as

$$\frac{d\sigma_c}{d\Omega} = \frac{r_0^2}{4}\left(\frac{\nu'}{\nu}\right)^2\left(\frac{\nu}{\nu'} + \frac{\nu'}{\nu} - 2 + 4\cos^2\theta\right) \text{cm}^2\,\text{sr}^{-1}\,\text{electron}^{-1} \quad (1.33)$$

where ν' and ν are the corresponding photon frequencies and θ is the angle between the electric vectors of the two photons.

When the incoming photon is unpolarised, the corresponding expression is now, θ being the angle of scatter:

$$\frac{d\sigma_c}{d\Omega} = \frac{r_0^2}{2}\left(\frac{\nu'}{\nu}\right)^2\left(\frac{\nu}{\nu'} + \frac{\nu'}{\nu} - \sin^2\theta\right) \text{cm}^2\,\text{sr}^{-1}\,\text{electron}^{-1} \quad (1.34)$$

As is well known, the cross sections reduce to the Thomson cross section $(8\pi r_0^2/3)$ in the limit of very low frequency and when integrated over the whole solid angle. At high photon energies the fall off in $d\sigma_c/d\Omega$ with increasing ν (Klein–Nishina formula) is apparent.

Compton telescopes using the above process have been used quite

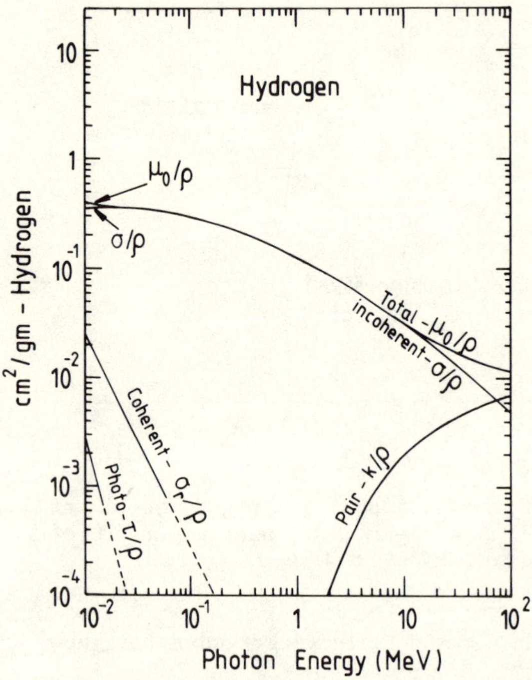

Figure 1.10. Mass attenuation coefficients for γ-rays in hydrogen for the more important processes (after Chupp 1976). The quantities μ, σ, K and τ are the linear attenuation coefficients so that each, divided by density, ρ, is the mass attenuation coefficient.

extensively in γ-ray experiments, the telescopes being carried aloft by balloons. More sophisticated versions in which photons interact twice ('double-Compton' telescopes) are being used increasingly.

Figure 1.10 gives the corresponding mass attenuation coefficient for Compton scattering in hydrogen, and also for the other relevant processes. As an example we can consider the standard 10 kpc of ISM of density 1 atom cm^{-3}. At 1 MeV the mass attenuation coefficient is 0.13 cm^2 g^{-1}, i.e. a mean free path of $\approx$8 g cm^{-2} hydrogen which, converting to units of length, is $8 \times (1.7 \times 10^{-24})^{-1}$ cm, i.e. $\approx$2 Mpc. The scattering in 10 kpc is thus quite negligible. The same situation does not apply in dense plasmas, however, and here Compton scattering can be quite important.

Pair production. The threshold for this process is $2mc^2(1 + m/M)$, where M is the mass of the particle (usually a nucleus) in whose Coulomb field the pair of particles, each of mass m, is being produced. For creation of electrons on a nucleus the threshold is $2m_ec^2 = 1.022$ MeV, and for creation of electrons on an electron (a rare phenomenon) the threshold rises to $4m_ec^2 = 2.044$ MeV.

Straightforward forms can be given for the pair production cross sections in the limiting cases of no atomic screening and complete screening.

For no screening (relevant to $1 \ll h\nu/m_ec^2 \ll 1/\alpha Z^{1/3}$),

$$\sigma_{pp} = \bar{\sigma} Z^2 \left(\frac{28}{9} \log \frac{2h\nu}{m_e c^2} - \frac{218}{27} \right) \text{cm}^2 \text{ atom}^{-1} \tag{1.35}$$

following Marmier and Sheldon (1969).

When $h\nu/m_ec^2 \gg 1/\alpha Z^{1/3}$, there is the condition of complete screening and the increase of σ_{pp} with ν is suppressed. The form is then

$$\sigma_{pp} = \bar{\sigma} Z^2 \left(\frac{28}{9} \log \frac{183}{Z^{1/3}} - \frac{2}{27} \right) \text{cm}^2 \text{ atom}^{-1} \tag{1.36}$$

The form of the corresponding mass attenuation coefficient is given in Figure 1.10; above 100 MeV pair production is seen to dominate. As with the Compton effect, losses in the general ISM are negligible, but in dense plasmas, such as might be found near black holes, the process can become important.

1.3.3 Gamma–gamma interactions

A γ-ray of energy ε_1 colliding with another of energy ε_2 gives rise to a pair of particles, each of mass m, if ε_1 is greater than a threshold value, ε_t, given by

$$\varepsilon_t = \frac{2m^2c^4}{\varepsilon_2(1 - \cos\theta)} \tag{1.37}$$

where θ is the angle between the photon trajectories. Restricting attention to electrons and the case of head-on collisions, we have

$$\varepsilon_t = 0.26 \times 10^{12}/\varepsilon_2$$

where the energies are in eV.

Considering now collisions with astronomically important radiation fields, we have (with their mean energies in brackets)

2.7 K radiation	(6×10^{-4} eV) $\varepsilon_t \approx 4 \times 10^{14}$ eV
starlight	(2 eV), $\varepsilon_t \approx 10^{11}$ eV
X-rays	(1 keV), $\varepsilon_t \approx 3 \times 10^{8}$ eV

The corresponding cross section for the production of pairs of particles each of mass m in head-on collisions is

$$\sigma_1 = \pi r_e^2\left(\frac{m}{\omega}\right)^2\left[2\ln\left(\frac{2\omega}{m}\right) - 1\right] \tag{1.38}$$

for $\omega \gg m$, i.e. in the extreme relativistic region, and

$$\sigma_2 = \pi r_e^2\left(1 - \frac{m^2}{\omega^2}\right)^{1/2} \tag{1.39}$$

for ω close to m, i.e. in the classical region. In the expressions $\omega = (\varepsilon_1\varepsilon_2)^{1/2}$ and r_e is the classical electron radius. (For further details see Jauch and Rohrlich (1955).)

Although the cross section is not small, photon–photon collisions are not generally frequent because the spatial density of target photons is not large. Nevertheless, even with low photon densities, such as encountered with the various extragalactic photon fields – $\approx 400\ \text{cm}^{-3}$ for the 2.7 K radiation and $\approx 5 \times 10^{-3}\ \text{cm}^{-3}$ for extragalactic starlight with an energy density of $\approx 10^{-2}\ \text{eV cm}^{-3}$ – the ensuing mean free paths for collision are such as to lead to a significant attenuation over typical extragalactic distances. Figure 1.11 shows the interaction lengths derived using expressions of the form given above but allowing for the geometry of the collisions and including the energy distribution of the target photons.

The extent to which γ–γ collisions are important in attenuating γ-rays can be seen by reference to Figure 1.11. On the Galactic scale the flux received from sources as far away as the Galactic Centre, or further, will be reduced for γ-rays in the region of 10^{15} eV. It will be seen later (Chapter 5) that the loss is serious for the source Cygnus X–3, the distance of which appears to be about 11 kpc.

At the extragalactic level, it will be seen that attenuation is important at all energies in the range considered in Figure 1.11. In fact, the interaction length for γ-rays on starlight photons becomes greater than the Hubble radius just below 10^{12} eV so that there is an effective transparency of the Universe for $E_\gamma < 10^{12}$ eV.

The behaviour of very-high energy photons, produced on a universal scale and propagating over extragalactic distances, is one of great interest and some complexity. Photon cascading occurs, with the electron pairs generated in γ–γ collisions themselves producing more photons by way of the inverse Compton interactions. Wdowczyk *et al.* (1972 and later papers) have considered this problem and derived the γ-ray energy spectrum to be expected from the injection of γ-rays of energy above

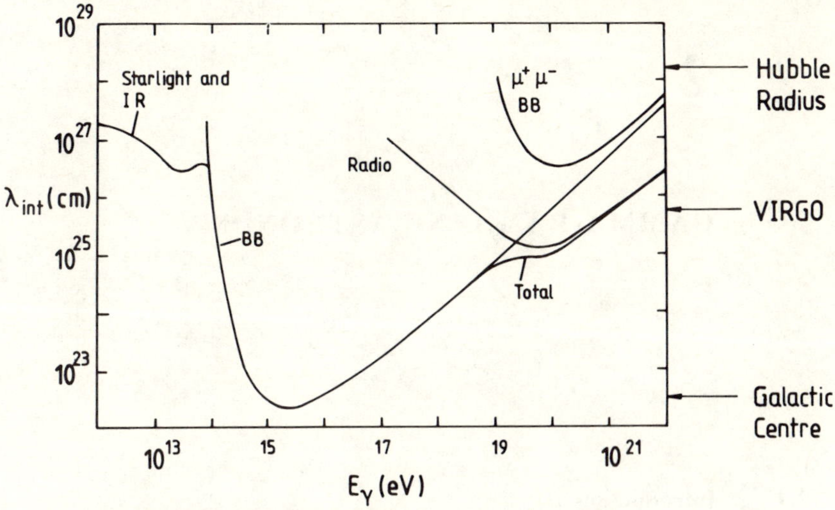

Figure 1.11. Interaction length of γ-rays on the various background radiation fields. Unless otherwise stated the process concerned is electron pair production. BB denotes the 2.7 K black body radiation (0.24 eV cm^{-3}, $\approx$400 photons cm^{-3}). Several important distances are indicated on the right-hand side; 'VIRGO' denotes the distance to the important cluster of galaxies at the centre of 'our' Supercluster. The results are from the work of Wdowczyk *et al.* (1972).

10^{19} eV such as might have come from the interaction of cosmic ray protons with the 2.7 K photons. The ensuing cascade gives a γ-ray cascade extending down to low energies but the intensity predicted, corresponding to a γ/proton ratio of $\approx 10^{-6}$ at $E_\gamma = 10^{12}$ eV, is too low to be detected yet.

The foregoing has been concerned mainly with interactions of γ-rays and the background photons. In the immediate vicinity of various exotic objects the X-ray and harder photon fluxes can be so high that γ-rays in the GeV and even MeV regions are seriously affected. The steepening of the energy spectrum of γ-rays from quasars at about 2 MeV may be due to this cause (see Chapter 4).

2

GAMMA-RAY LINE ASTRONOMY

2.1 Introduction

The spectroscopy of γ-ray astronomy is, understandably, an area where important advances are to be expected, an expectation born of similar previous experience with other regions of the electromagnetic spectrum. Technical difficulties are considerable at present, however, due to low line fluxes agggravated by serious background problems; nevertheless, a promising start has been made and several interesting observations have already appeared.

As with astronomy in general, a distinction can be made between observations of 'discrete' objects (such as stars, supernovae, other galaxies etc.) and signals from more extended regions, in particular the interstellar medium (ISM).

In the first category, γ-ray lines from the sun – due to energetic protons and heavier nuclei interacting with the solar atmosphere – provide interesting and important information about a variety of solar phenomena. This subject of solar γ-ray spectroscopy is distant from the main stream of topics discussed here and the reader is directed to a number of useful reviews by Ramaty and Lingenfelter (1981), Trombka and Fichtel (1982), and the books by Chupp (1976) and Hillier (1984).

In the non-solar region, which is of main concern here, only three γ-ray lines have been detected from non-transient celestial sources so far. These are the line at 1809 keV from the Galactic Equatorial Plane, the line at 511 keV from the Galactic Centre and the one at 1369 keV from the object SS 433; these will be described in Sections 2.2, 2.3 and 2.4, respectively. We will then briefly present in Section 2.5 the searches made for γ-ray lines from extragalactic objects. Finally, in Section 2.6, there is a summary of the results obtained so far and a résumé of their implications.

2.2 Lines from the interstellar medium

2.2.1 Expectations

Chapter 1 dealt with the manner in which ISM lines are expected to be produced; here we give more details. Ramaty and Lingenfelter (1979) have made extended calculations of the lines expected from cosmic ray interactions in the ISM, with the result shown in Figure 2.1. Two classes of lines can be identified: sharp and broad. The sharp lines come from interactions with interstellar dust, the recoil nuclei in the dust grains coming to rest before de-exciting. The broader lines are due to the ISM in gaseous form; to give a perspective, it is likely that about 15 % of carbon and oxygen, for example, are in the gaseous ISM. The broad lines will also have a contribution from cosmic ray nuclei ($Z > 1$) rendered radioactive by collisions with the nuclei of the ISM.

Sharp lines from the dust–grain phase of the ISM are expected for the common elements in the ISM: oxygen (^{16}O), magnesium (^{24}Mg) and iron (^{54}Fe, ^{55}Fe, ^{56}Fe), as can be seen in Figure 2.1. Of these lines the ^{16}O line at 6.129 MeV is most conspicuous in view of the lack of other lines in its immediate neighbourhood.

It should be stressed that the estimates in Figure 2.1 are necessarily approximate because of uncertainty in both the distribution of cosmic ray intensity and in the heavy element abundances along the line of sight. The topic of cosmic ray intensity variations in the Galaxy is taken up in Chapter 4.

The positron annihilation line is also seen to be quite marked, the positrons having come from the $\pi^+ \rightarrow \mu^+ \rightarrow e^+$ chain and having slowed down to very low velocities before annihilating (the e^+e^- cross section for annihilation is high only at such low velocities). The predicted strength of the 511 keV line is $\approx 7 \times 10^{-5}$ cm^{-1} s^{-1} rad^{-1}, i.e. about 3×10^{-5} cm^{-2} s^{-1} for a detector of resolution 0.2 sr (FWHM = 30°) centred on the Galactic Centre.

The possibility of detecting γ-ray lines from supernova remnants (SNR) is one of great potential interest because of its relevance to the physics of supernova (SN) explosions and to the (likely) demonstration of continuing nucleosynthesis. The importance of this field seems to have been recognised first by Clayton *et al.* (1969); later calculations have been made by a variety of authors (e.g. Clayton 1973, Lingenfelter and Ramaty

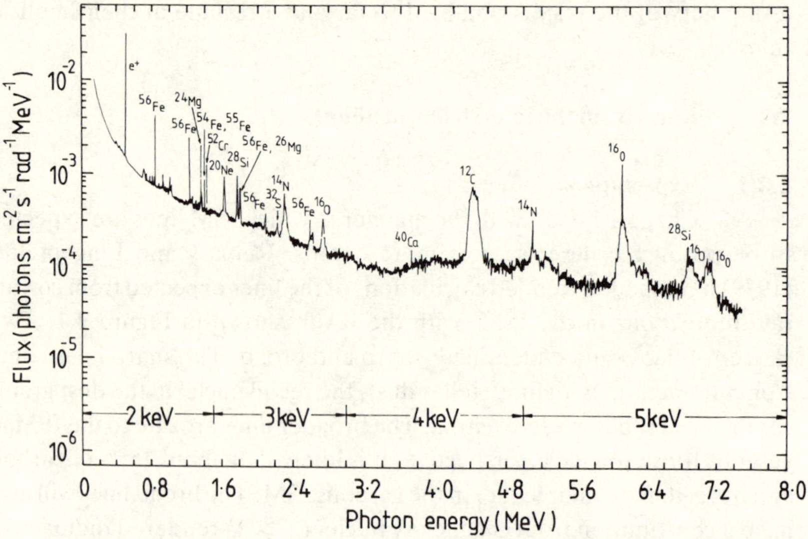

Figure 2.1. The expected γ-ray spectrum in the general direction of the Galactic Centre arising from cosmic ray interactions in the interstellar medium (after Ramaty and Lingenfelter 1979). The ranges marked '2 keV, 3 keV . . .' refer to the instrumental line widths assumed in the calculations and used to derive the indicated line fluxes. For example, the ^{16}O line at 6.13 MeV plotted with a flux, above background, of 1.2×10^{-3} cm^{-2} s^{-1} rad^{-1} MeV^{-1} would have a flux of 5×10^{-3} times this value, i.e. 6.0×10^{-6} cm^{-2} s^{-1} rad^{-1} when seen with a detector having a FWHM of 5 keV.

1978, Woosley *et al.* 1981). Short-lived radioactive nuclei will generate γ-rays which are unlikely to escape from the high density material around a very young SN but sufficient amounts of isotopes with longer mean life will survive until the SNR density is low enough to allow penetration. Nuclei with lifetimes longer than a few times 10^5 y will survive so long that the remnant will have merged into the general ISM, and after subsequent mixing these radioactive nuclei will add their γ-rays to the reservoir from cosmic ray interactions. The isotope ^{26}Al (mean life = 1.04×10^6 y) is of particular interest in this context because it is almost certainly produced largely in SN and novae explosions and its study can allow inferences as to how much nucleosynthesis takes place in these explosive cosmic occurrences.

Ramaty and Lingenfelter (1979) give data (Table 2.1) for the more important isotopic decay chains associated with the nucleosynthesis reactions occurring in supernovae. The ensuing lines are indicated in Figure 2.2, superimposed on the 'ordinary' ISM lines; also shown in the

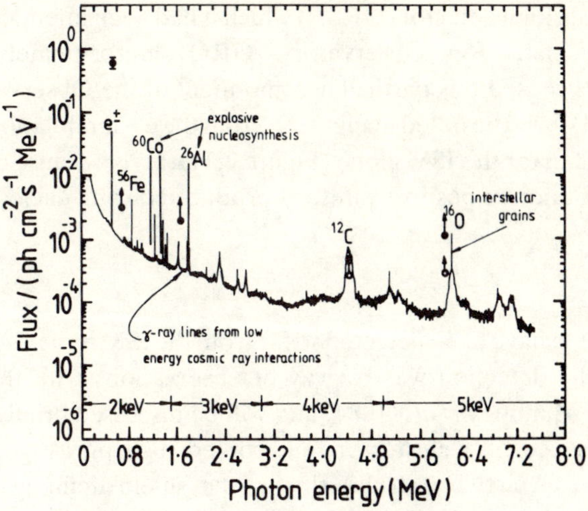

Figure 2.2. Diffuse γ-ray line emission from the interstellar medium, including contributions from explosive nucleosynthesis (after Ramaty and Lingenfelter 1979). The points with upward arrows represent the sensitivities expected for an instrument which was earlier planned to be part of the forthcoming Gamma Ray Observatory, as described in the text.

Table 2.1. *The decay chains from nucleosynthesis in supernovae (after Ramaty and Lingenfelter 1979)*

Decay chain	Mean life (y)	$Q/Q(^{56}\mathrm{Ni})$	Energy	Positrons or photons per disintegration
^{56}Ni → ^{56}Co → ^{56}Fe	0.31	1	e$^+$	0.2
			0.847 MeV	1
			1.238 MeV	0.7
^{57}Co → ^{57}Fe	1.1	2×10^{-2}	0.122 MeV	0.88
			0.014 MeV	0.88
^{22}Na → ^{22}Ne	3.8	5×10^{-3}	e$^+$	0.9
			1.275 MeV	1
^{44}Ti → ^{44}Sc → ^{44}Ca	68	2×10^{-3}	e$^+$	0.94
			1.156 MeV	1
			0.078 MeV	1
			0.068 MeV	1
^{60}Fe → ^{60}Co → ^{60}Ni	4.3×10^5	1.5×10^{-4}	1.322 MeV	1
			1.173 MeV	1
			0.059 MeV	1
^{26}Al → ^{26}Mg	1.1×10^6	1.5×10^{-4}	e$^+$	0.85
			1.809 MeV	1

figure are the flux limits for a detector (GRSE) which it had been intended to fly aboard the Gamma Ray Observatory (GRO); unfortunately, shortage of funds has caused this particular component of the Observatory to be abandoned. It will be noted in Figure 2.2 that the e^+ annihilation line is higher in flux than for the ISM alone (Figure 2.1); this is because of the additional contribution of positron emitters produced during nucleosynthesis (Table 2.1).

2.2.2 Observations

The only line to have been detected so far from the ISM is the ^{26}Al line at 1809 keV. The detection was by way of observations with the HEAO-3 satellite by Mahoney *et al.* (1984) and will be discussed briefly. The instrument consisted of a cluster of four cooled high purity germanium detectors in an active CsI(Na) shield, the shield defining a geometrical aperture of about 30° FWHM. Mahoney *et al.* (1980) have described the instrumentation in detail.

HEAO-3 was launched on September 20, 1979, and although the instruments worked well, a build up of radiation contamination occurred due to cosmic ray bombardment. Of specific importance to the search for narrow lines was the instrumental width of the lines – this increased from 3.5 keV (FWHM) just after launch to about 19 keV in the late spring of 1980 (Mahoney *et al.* 1980). A preliminary search (Mahoney *et al.* 1982) gave an indication of the ^{26}Al line at the 2.6σ level, and more recent analysis has confirmed the result at a higher level of significance. The analysis was performed with considerable difficulty because of the low signal strength, high background rate and poor angular resolution. Unfortunately, it was not possible to study the longitude distribution, which would have enabled the mode of production of the ^{26}Al line to be examined in some detail. Instead, a model longitude distribution was adopted and the appropriate average value fitted. The model taken assumed that the longitudinal dependence was the same as that of diffuse γ-rays (above 70 MeV) as measured by the SAS II and COS B satellites (see Chapter 4 for a full description of these results). Essentially, this distribution has a ratio of ~4 : 1 for the Galactic Centre (GC) to Galactic Anti-Centre directions, the width of the broad excess towards the GC is $\approx \pm 60°$ in longitude and the flux towards $l = 0°$ is about twice the mean flux.

The result of the fitting is to give a line flux of the ^{26}Al γ-rays towards the GC of $(4.8 \pm 1.0) \times 10^{-4}$ cm^{-2} s^{-1} rad^{-1} with a line width of <3 keV (FWHM). Figure 2.3 summarises the results.

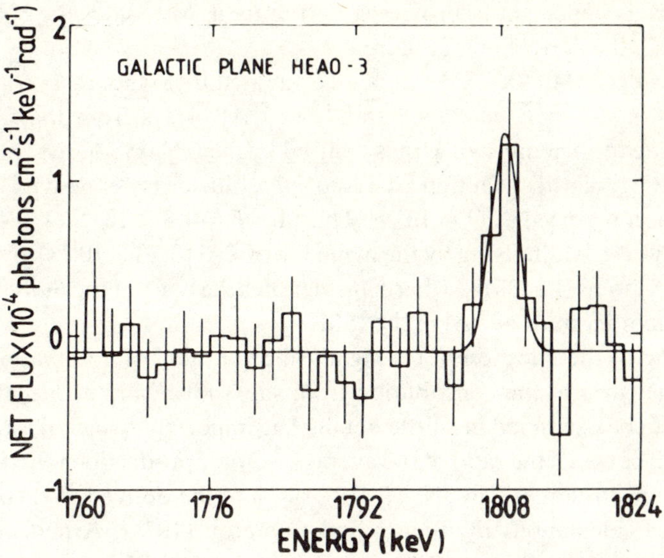

Figure 2.3. HEAO-3 results on the ^{26}Al line (Mahoney *et al.* (1984). The data refer to the total net Galactic Plane emission near 1809 keV normalised to the direction of the Galactic Centre. The data are those of highest quality and were taken during autumn, 1979. The channel width is 2 keV and the solid curve is the best fit for a background plus an unresolved line at 1809 keV. The fit gives a line with an energy of 1808.49 ± 0.41 keV consistent with the expected 1808.65 ± 0.07 keV emission from ^{26}Al at rest. The width in the fit is consistent with the instrumental resolution and gives a 1σ upper limit of 3 keV FWHM for the intrinsic width of the line.

We shall first examine the significance of the line width. The lifetime of ^{26}Al is such that most of its nuclei will have reached equilibrium with the gas in the ISM and thus have velocities <10 km s^{-1}. A bigger contribution to velocity – and thus Doppler broadening – is from Galactic rotation and this corresponds to a velocity $v \lesssim 250$ km s^{-1}; the fractional line width will therefore be less than, approximately, $v/c = 2.5 \times 10^7/3 \times 10^{10}$, i.e. $<10^{-3}$, or, equivalently <2 keV for the 1809 keV line. There is seen to be no inconsistency with the results of Mahoney *et al.* (1982).

Concerning the predicted flux, Figure 2.2 indicates a value of $\approx 1.5 \times 10^{-2}$ cm^{-2} s^{-1} MeV^{-1} which, for a line width of 3 keV (the value assumed in drawing the spectrum), translates to 4.5×10^{-5} cm^{-2} s^{-1} rad^{-1} (allowing for the fact that the flux is roughly constant towards the GC over about 1 radian). The measured value of 4.8×10^{-4} cm^{-2} s^{-1} rad^{-1} is thus an order of magnitude higher than expected. It seems likely, then, that some other source is important in providing ^{26}Al in the ISM besides the SNR considered in the derivation of Figure 2.2

The mass of ^{26}Al in the ISM has been estimated by Mahoney *et al.* as follows. With a flux $\bar{F}$ (cm^{-2} s^{-1} rad^{-1}) of ^{26}Al γ-rays from the GC direction, the total number of γ-rays emitted by the Galaxy follows from the assumed radial distribution of radioactive nuclei ($\propto$ emissivity of medium energy γ-rays) as $Q = 10^{46}\bar{F}$. Thus, for $\bar{F} = 4.8 \times 10^{-4}$, $Q \approx 4.8 \times 10^{42}$ γ-rays s^{-1}. Multiplying by the mean life of ^{26}Al (3.28×10^{13} s) gives $\approx 1.6 \times 10^{56}$ atoms (i.e. $3M_\odot$, where $M_\odot$ denotes the mass of the Sun) for the contemporary mass of ^{26}Al in the ISM.

Returning to the question of the significance of the ^{26}Al observation, and the relevance to the contribution from supernovae and novae, the situation can be examined in a little detail. The important question is the relationship between the ratio of the average isotopic production ratio for explosive nucleosynthesis in SN: $(P_{26}/P_{27})_{SN}$, and the derived ^{26}Al content. Model calculations by Truran and Cameron (1978), Arnett and Wefel (1978), Arnould *et al.* (1980), and Woosley and Weaver (1980), give values in the range $(0.4–2.0) \times 10^{-3}$ for this ratio, and this range appears to be acceptable. However, the conversion depends on the ratio of the average Galactic metallicity to that of solar system matter $(A_G/A_\odot)$ and on the appropriate Galactic mass to adopt (M_G). Mahoney *et al.* (1984) adopting $A_G/A_\odot = 2.3$ and $M_G = 2 \times 10^{11} M_\odot$ use the measured ^{26}Al content to derive $(P_{26}/P_{27}) \approx 4 \times 10^{-3}$, not far from expectation. Clayton (as described by Mahoney *et al.* 1984) has pointed out that the values adopted for $A_G/A_\odot$ and M_G were inappropriate, however, and using the values ~ 1 and $7 \times 10^{10} M_\odot$, respectively, appropriate to Population I star material, which is the product of SN, P_{26}/P_{27} moves up to 30×10^{-3}, a factor of at least 15 above the expected value. This conclusion is consistent with that following from the calculations of Ramaty and Lingenfelter (1979) and given in Figure 2.2. It does appear likely therefore that there is an excess of ^{26}Al over expectation from SN alone by a factor of about ten.

Mahoney *et al.* (1984) have addressed the question of other sources, starting with novae. They refer to the calculations of Hillebrandt and Thielemann (1982) which yield an expected mass of $\sim 1M_\odot$ of ^{26}Al from novae. The total expectation is now $1.3M_\odot$ (SN $0.3M_\odot$, novae $1M_\odot$), to be compared with the observed value of $3M_\odot$. It is not clear whether the difference is significant, bearing in mind the problems referred to earlier concerning the derivation of the GC γ-ray line flux; nevertheless suggestions have been made that red giants or massive main sequence stars may also be contributing appreciably to the ^{26}Al budget.

2.3 The 511 keV line from the Galactic Centre

2.3.1 The observations

The progress of research on the positron annihilation line from the GC epitomises the whole subject of γ-ray astronomy. Early balloon observations gave tantalising evidence for a significant flux, but high backgrounds and an apparent variability of signal caused much scepticism. However, following more recent satellite observations it is generally agreed that there *is* a significant flux and that it *does* vary somewhat with time.

The first observation of a line in the region of 511 keV seems to have been made by Johnson and Haymes (1973) – the energy in fact being reported as $\approx$476 keV. Two years later, in a repeat experiment, Haymes *et al.* (1975) found a line again, although this time at $\approx$530 keV. Technical improvements – principally the use of a cooled Ge(Li) detector – by Leventhal *et al.* (1978) led to the detection of a line feature at 511 keV, with a resolution of 3.2 keV, the flux being 1.2×10^{-3} cm^{-2} s^{-1}. More recent observations by Leventhal *et al.* (1980) and Albernhe *et al.* (1981) have confirmed the existence of the line, but the most significant results have come from the Jet Propulsion Laboratory's γ-ray spectrometer on the HEAO-3 satellite (Riegler *et al.* 1981). The satellite data have given an energy of 510.9 ± 0.25 keV and a width less than 2.5 keV for the line. The exact location of the source is still somewhat uncertain; however, the centroid of the source is known to lie within a few degrees of the GC but its width can be quite large. Alternatively, there might be a number of separated but unresolved discrete sources (the FWHM of the detector is 30° and all that can be said is that the extent is much less than this).

2.3.2 The main features of the 511 keV line

Figure 2.4 shows the signal from a 1979 Galactic Plane scan by HEAO-3 (Riegler *et al.* 1981); there can be no doubt about the statistical significance of the GC line.

There are two remarkable features about this line: its magnitude and its time variability. Although the cosmic ray intensity at the GC is probably higher than locally, and also there is an abundance of target gas (see Bhat *et al.* 1985), the contribution of cosmic ray-generated positrons appears to

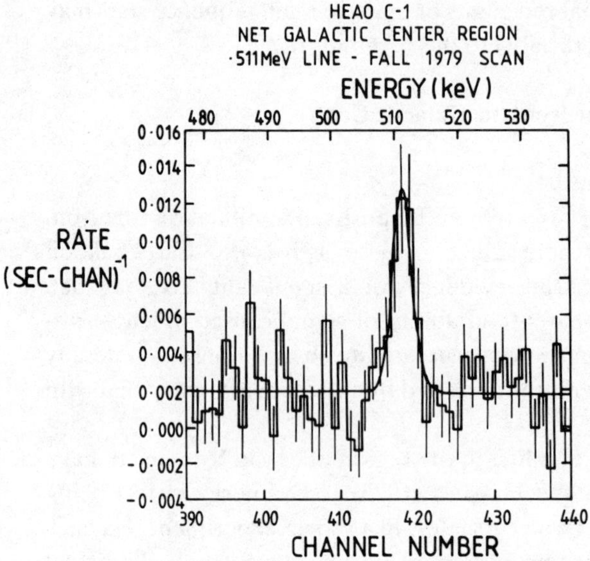

Figure 2.4. HEAO-3 observations of the 511 keV positron annihilation line from the 'Galactic Centre' (Riegler *et al.* 1981). The observations were taken in the autumn of 1979. The emission is centred on 510.90 ± 0.25 keV. The observed width is (3.13 ± 0.57) keV, to be compared with an instrumental resolution of 2.72 keV; the corresponding intrinsic width is 1.6 (+0.9 − 1.6) keV (all the widths are FWHM).

be negligible. This fact follows from the lack of observation of other cosmic ray-induced nuclear lines in the 1–2 MeV region from Mg, Si and Fe (Ramaty and Lingenfelter 1979) and the strong 4.4 MeV line emission from the de-excitation of ^{12}C. Matteson *et al.* (1979) estimate that no more than about 20 % of all the 511 keV line intensity could be due to cosmic rays of energy below about 100 MeV per nucleon. The preceding lines would have come from interactions not sufficiently energetic to produce pions. However, if positrons from the $\pi \rightarrow \mu \rightarrow e$ chain were to be responsible for the observed line the ensuing γ-rays from the associated π^0 decays would be very intense. There is, in fact, evidence for a peak from the GC direction in the 100 MeV region (as will be described later), but its flux is too low by a factor of at least 100 to be consistent with the 511 keV line if the γ-ray lines were produced by the annihilations of positrons resulting from the $\pi^+ \rightarrow \mu^+ \rightarrow e^+$ chain. Reverting to the actual magnitude of the line flux as observed in the 1979 observation (Figure 2.4) the corresponding source luminosity is $L \approx 2 \times$

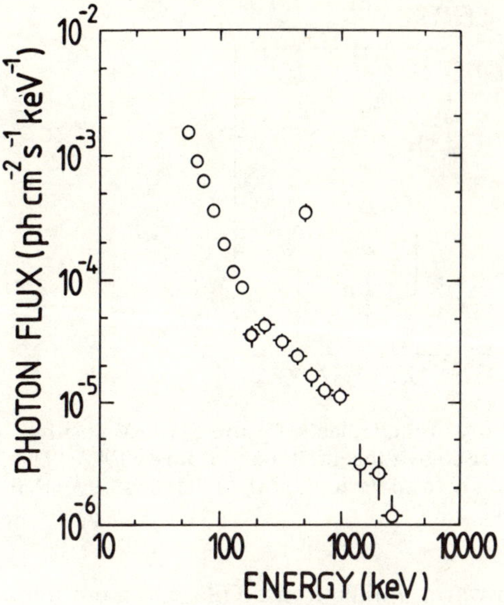

Figure 2.5. HEAO-3 observations of the Galactic Centre region for a 20-day period between September 24 and October 13, 1979 (Riegler *et al*. 1983). The 511 keV point is indicated for a bandwidth of 6 keV. The significance of the lines is described in the text.

10^{37} erg s^{-1} and the positron annihilation rate is $\approx 10^{43}$ s^{-1} (adopting a distance of 10 kpc for Sun–Galactic Centre).

At this stage it is appropriate to mention that later analysis by Riegler *et al*. (1983) has indicated a continuum both below and above 511 keV from the GC direction. Figure 2.5 shows the situation. The spectral characteristics are important from the standpoint of determining the relative fractions of γ-rays coming from direct e^+e^- annihilation. Although singlet state positronium yields two photons of energy 511 keV, the triplet state (formed three times as often) yields three photons of variable energy below 511 keV. It is possible to determine the relative fraction, f, of the 511 keV line, which has come by way of positronium compared with direct annihilation, by fits to the points in Figure 2.5. Riegler *et al*. derive $f = 0.58 \pm 0.17$ at the 90% confidence level, a value somewhat smaller than $f = 0.9$ quoted in the earlier experiment of Leventhal *et al*. (1978) but not inconsistent. The question of the expected value of f has been considered by Bussard *et al*. (1979) and it seems that the observed values are in the expected region in the sense that both the line width and the f-values would be consistent with positrons annihilating

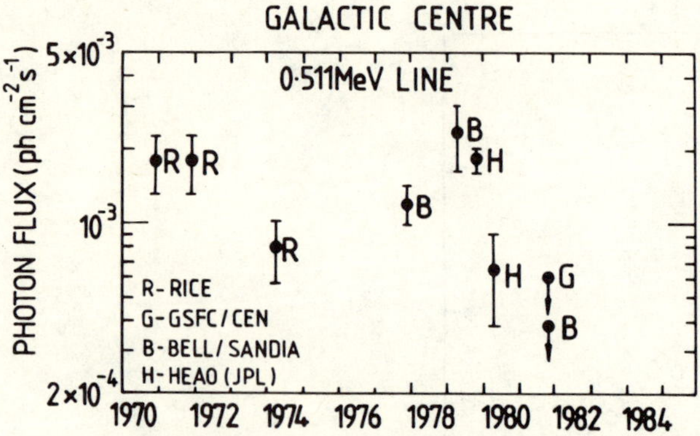

Figure 2.6. Time variability of the Galactic Centre 511 keV positron annihilation line (summarised by Lingenfelter and Ramaty 1982). The two upper limits G and B are from Paciesas *et al.* (1982) and Leventhal *et al.* (1982), respectively.

and forming positronium in a warm, mildly ionised plasma, a not unexpected environment for the GC region.

Turning to the time variability, Figure 2.6 shows the situation. The variability, which seems quite genuine, suggests that the object is confined in space, presumably to less than about a light year (or, more precisely, the e^+e^- annihilation region is localised within this magnitude). It is interesting to note that a somewhat similar variability has been noted in the hard X-ray region.

The models put forward to account for the above features will be discussed in the next section.

2.3.3 Models for the origin of the 511 keV line

A comprehensive study of models has been made by Lingenfelter and Ramaty (1982) and this study will be followed here.

The most important experimental features of the lines are those listed already: flux, width and variability. As remarked in the preceding section, the width is related to the properties of the gas in which the line originates; analysis indicates an ionisation level of at least 10 % and a temperature below 5×10^4 K. The time variability suggests a linear dimension for the emission region of less than about 0.3 pc. It is interesting to note that such conditions do appear to be present in the warm clouds and other small concentrated infra-red sources within about a parsec of the GC.

The crucial question of the source of the positrons will now be addressed. The strongest constraints on the e^+ production process appear to be set by the limit on the accompanying continuum emission at energies above 511 keV. The process requires a very high efficiency such that about half of the total radiated energy above 511 keV emerges as e^+e^- pairs, assuming, that is, that the positron production occurs on similar time scales to that of the observed γ-rays and that there is isotropy.

Lingenfelter and Ramaty strongly favour γ–γ collisions as the source of the electron–positron pairs, the source being required to have a dimension of less than about 10^8 cm in order to meet the luminosity requirement. Not surprisingly, a black hole is postulated as the ultimate 'engine'. An accreting black hole of quite low mass ($<10^3 M_\odot$) surrounded by a gas at MeV temperature, emitting intense bremsstrahlung, appears to be very reasonable. Such a mass for the black hole is much smaller than that suggested for several other phenomena, such as the possible presence, at 10^7 y intervals, of explosions in the Galactic Nucleus, explosions which might contribute significantly to the cosmic ray flux. The two hypotheses are not necessarily inconsistent, however, in that black hole characteristics are not well understood and, furthermore, the Galactic explosions are highly speculative.

2.4 Gamma-ray lines from other Galactic objects

2.4.1 The CRAB pulsar and nebula

Extensive measurements have been made of the CRAB region at all wavelengths, from long radio waves to ultra-high energy γ-rays. The situation with respect to γ-ray lines is not clear cut. There have been reports of line emission at about 73 keV (Ling *et al.* 1979), supposedly arising from cyclotron emission of electrons and positrons in the very strong magnetic field close to the pulsar, but other searches for this emission feature have been unsuccessful. Restricting attention to the higher energy region, which is the subject of the present work, the situation is as follows. Leventhal *et al.* (1977) reported a line at 400 keV of width less than 3 keV (FWHM) and flux $(2.24 \pm 0.65) \times 10^{-3}\ \mathrm{cm^{-2}\ s^{-1}}$, from a balloon experiment on May 10–11, 1976. This flux was higher than the 3σ upper limit of $1.7 \times 10^{-3}\ \mathrm{cm^{-2}\ s^{-1}}$ placed on the line some two years earlier by Ling *et al.* (1977). Similarly, more recent observations by Hemeury *et al.* (1983), on September 25, 1980, placed a 3σ upper limit to the line of $1.7 \times 10^{-3}\ \mathrm{cm^{-2}\ s^{-1}}$.

Most recently, Mahoney *et al.* (1984) used the HEAO-3 detector to search for such lines. Although during normal operations the spin axis of the 'observatory' was pointed towards the sun, during two periods in 1979

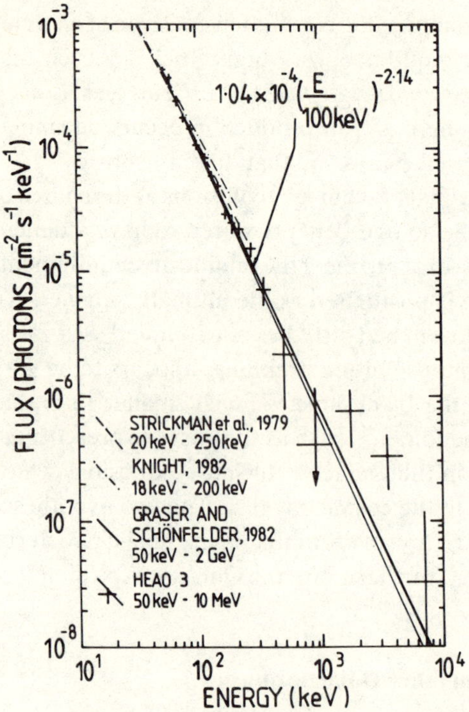

Figure 2.7. Energy spectrum of the CRAB pulsar averaged over the whole phase, from the HEAO-3 experiment and other observations (after Mahoney *et al*. 1984). Previous estimates are also shown. The heavy line is the best power law fit to the HEAO-3 data.

and 1980 its spin axis was oriented towards a Galactic Pole so that the C-1 instrument (γ-ray spectroscopy experiment) which was mounted perpendicular to the spin axis, scanned the Galactic Equatorial Plane. The CRAB passed within 20° of the 'look axis' for 18 days in the autumn of 1979 (from September 23) and for 41 days in the spring of 1980 (from February 24). The data accrued therefore represent the best available to date.

The results showed no evidence for line emission from the CRAB pulsar at an energy above 100 keV at the 3σ level of about 2×10^{-4} cm^{-2} s^{-1}, the value corresponding to an assumed incident line with an intrinsic width less than the energy resolution of the instrument. This upper limit is seen to be a factor of ten less than the claimed 1976 value (Leventhal *et al*. 1977) and we are left with the standard problem of whether the observation of Leventhal *et al.* was genuine or whether the line output is highly time variable. The answer cannot be given yet.

Although not strictly relevant to the line problem, mention should be made of the rather precise measurement of the pulsed continuum γ-ray flux from the CRAB pulsar with the HEAO-3 experiment. The phase-averaged spectrum was measured from 50 keV to 10 MeV and found to be consistent with a single power law, in agreement both in spectral index and intensity with the earlier analysis of Graser and Schonfelder (1982). Figure 2.7 shows the spectrum.

The 'light curve' of the γ-ray emission in this energy band is very similar to that at higher γ-ray energies, namely two main peaks of approximately equal intensity connected by strong interpeak emission. There was no significant change in the profile over the six months separating the observations, unlike the situation at higher energies (to be described in Chapter 4) where some variability has been reported in the observed light curve.

2.4.2 SS 433

SS 433 is a remarkable object, some 5.5 kpc away, which appears to comprise oppositely directed jets of relativistic material ($v \approx 0.26c$; Abell and Margon 1979, Milgrom 1979), and it is an obvious candidate for γ-ray searches.

HEAO-3 (Lamb *et al.* 1983) searched for emission from the direction of SS 433 and found a signal which is present at the 6σ level. The line in question was at 1.5 MeV and appeared variable by a factor of three on a time scale of days during the 46-day observation period. Figure 2.8 shows the result. In addition, the authors also saw a line at 1.2 MeV. Although there is no doubt about the veracity of the signal, the identification with SS 433 is somewhat uncertain in view of the poor angular resolution of the detector (FWHM 30°). Nevertheless, we proceed with the assumed identification in the following discussion.

The line near 1.5 MeV was detected at a level of 2–4σ in each of a series of three intervals and it is the superposition of six such sets of data which gives rise to the results shown in Figure 2.8. Inspection of likely γ-ray line energies shows none at the energy detected, and Lamb *et al.* were thrown back on the hypothesis of red- or blue-shifted lines. Two stand out as candidates: a blue-shifted decay of the first excited state of ^{24}Mg at 1.369 MeV or a red-shifted decay of the first excited state of ^{28}Si at

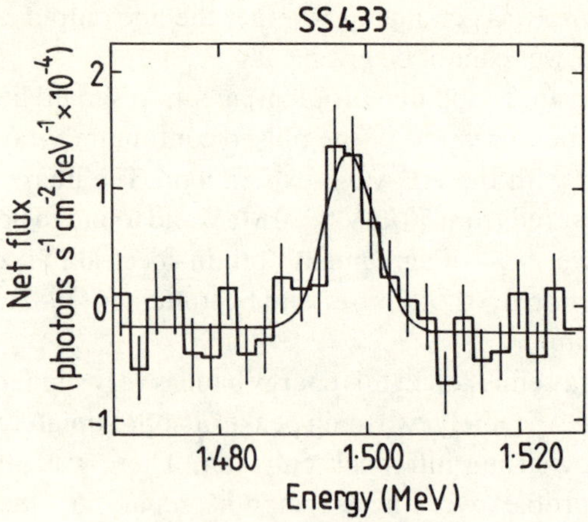

Figure 2.8. Averaged γ-ray flux from the region of SS433 within ± 30 keV of the 1.5 MeV feature, as detected by HEAO-3 (Lamb *et al.* 1983). The centre of the 'line' is at 1.497 MeV and the shape is 'consistent with a Gaussian of FWHM ≈ 9 keV'.

1.779 MeV. Of these the workers chose ^{24}Mg as the more likely emitting isotope since there is some evidence, too, for a red-shifted ^{24}Mg line at 1.2 MeV and this would be expected on the basis of the twin-jet model of SS 433. Although the precision is poor, there is also the suggestion of a change of line energy with time, such as would be expected from the model which explains the optical features, namely a precession of the axis of the jets with a 163.51-day period (Margon 1982). This feature increases confidence in the results somewhat. The authors found no evidence for the unshifted 1.369 MeV line and placed a 3σ upper limit at 1×10^{-3} photons cm^{-2} s^{-1} for this line.

Turning to the energy involved, this is considerable. If the lines correspond to near isotropic emission (likely even though the particles initiating the γ-rays are collimated in two jets), a γ-ray line luminosity of 2×10^{37} erg s^{-1} results. This is about 250 times that emitted in the X-ray region from 2 to 10 keV.

The theoretical estimates of the particle beam kinetic energy are in the range 10^{38}–10^{41} erg s^{-1} (Davidson and McCray 1980, Begelman *et al.* 1980) so that the γ-rays may well represent a significant fraction of the beam energy.

The interesting topic of γ-ray lines from *transient* sources is taken up in Chapter 3.

2.5 Gamma-ray lines from extragalactic objects

2.5.1 A search for lines from near-normal galaxies

The HEAO-3 satellite has been used by Marscher *et al.* (1984) to search for the 511 keV annihilation line from what the authors term 'mildly active galaxies', in that these galaxies have somewhat active central regions similar to our own. On the basis of a simple model for the γ-ray line emission mechanism from the GC in which γ-ray emission is related to radio emission, these workers estimate that five of the six galaxies studied should have detectable fluxes (see Table 2.2). In actual practice none was detected, and for these five the 98% upper limit was less than the expected flux.

The obvious conclusion to be drawn is that the model is inapplicable. A number of possibilities arise. For example, the GC measurements may refer to an unusually 'high' period in its certainly variable output characteristic. Another obvious possibility is that the GC γ-rays are beamed. A further idea is that the 'mildly active' galaxies do in fact have sufficient activity to eject part of the target gas from the region where the positrons would be expected to annihilate. Further GC observations should help test at least the first hypothesis.

2.5.2 Gamma-ray lines from Centaurus A

γ-ray lines at 1.6 and 4.4 MeV have been reported by Hall *et al.* (1976), although the 511 keV line was not seen. Fabian *et al.* (1976) have put forward a possible explanation in terms of quasi-thermonuclear reactions near a black hole, where the absence of the 511 keV line could

Table 2.2. *Expected flux and observed limits to the 511 keV line, from the work of Marscher* et al. *(1984)*

Expectation is based on a mean Galactic Centre flux of 1.3×10^{-3} cm^{-2} s^{-1} and the assumption that the line strength is proportional to the radio flux.

Galaxy	F_{511} (expected) (10^{-4} photons cm^{-2} s^{-1})	F_{511} (observed upper limit) (10^{-4} photons cm^{-2} s^{-1})
M81/M82	5.3	<2.4
NGC 4278	9.2	<3.5
M104	1.4	<3.9
NGC 6500	3.7	<2.8
NGC 2911	3.7	<2.2
NGC 262	4.6	<2.5

be due to the smearing effect of large Doppler broadening in the hot accretion gas. Clearly it would be wise to have confirmation of the lines before pushing conjecture too far.

2.5.3 Gamma-ray lines from other galaxies

The Seyfert galaxy NGC 4151 has been detected in the MeV region by a number of authors. Although lines have not been claimed yet, the steepening of the γ-ray spectrum above a few MeV (see Dean and Ramsden 1981 for a summary) – which may well be a feature of Seyferts – *could* be due to a number of unresolved lines in the MeV region thereby accentuating what would otherwise have been a gradual change of slope. This feature is relevant to the interpretation of the so-called diffuse flux of extragalactic X-rays and γ-rays (see Chapter 4).

Future plans for line measurements are considered in Chapter 4, and we shall see that the GRO has potential here.

2.6 Summary

It will be clear from what has been written that γ-ray line astronomy has started in earnest with a number of interesting and widely different phenomena.

At the lowest energy, there is the detection of the e^+e^- annihilation line from the GC region. Although the linear dimension of the 'source' cannot be determined experimentally, the time variability over a period of years strongly suggests that the source is 'small' ($\leqslant$1 parsec); the time variability is also of importance because of its relevance to the mechanism of line formation and of the origin of the parent positrons. An exotic 'engine' generating these particles is indicated.

Moving to the interstellar medium, a strong line at 1809 keV, presumably due to ^{26}Al, has been detected. Its strength appears to be excessive in comparison with expectation for the ejecta of SN. The much more common novae may be contributing significantly.

Lines have also been detected from discrete celestial objects, specifically the CRAB nebula and the remarkable object SS 433. Concerning the latter, there appears to be evidence for the Doppler shift in energy in line with the twin-jet model; this feature is an excellent example of how powerful γ-ray line studies will be for a whole collection of objects once measurements of adequate statistical precision are available.

3

GAMMA-RAY BURSTS

3.1 Introduction

The discovery of γ-ray bursts (GRB) was serendipitous as was that of pulsars discovered at about the same time. Pulsars were first detected in 1967 in an experiment designed to study interplanetary scintillation of compact radio sources, and the discovery paper (Hewish *et al.* 1968) was subsequently published; the first GRB was also seen in the year 1967 (although not reported until six years later; Strong *et al.* 1973) in a satellite-borne detector meant to monitor violations of the nuclear explosion test ban treaty. The publication of the discovery of GRB was first made in 1973 by Klebesadel *et al.* (1973). The detector comprised six caesium iodide scintillators, each of 10 cm^3, mounted on each of the four VELA series of satellites (5A, 5B, 6A and 6B), these vehicles being arranged nearly equally spaced in a circular orbit with a geocentric radius of $\sim 1.2 \times 10^5$ km. The detectors were sensitive to individual γ-rays in the approximate energy range 0.2–1.5 MeV and the detector efficiency ranged from 17–50%.* The scintillators had a passive shield around them; background γ-ray counting rates were routinely monitored. A statistically significant increase in the counting rates initiated the recording of discrete counts in a series of quasi-logarithmically increasing time intervals. The event time was also recorded. Data were telemetered down to the ground-based receiving stations. Unexpectedly large increases in the γ-ray counting rates that occurred nearly simultaneously in detectors on

* In many situations the term 'γ-rays' is applied only to quanta with energies greater than several hundred – perhaps 511 – keV. In the context of GRB, it is customary to designate all quanta with energies of some 20 keV and above as 'γ-rays'; for one finds a continuous spectrum of quanta all the way from 20 keV to MeV energies in the bursts.

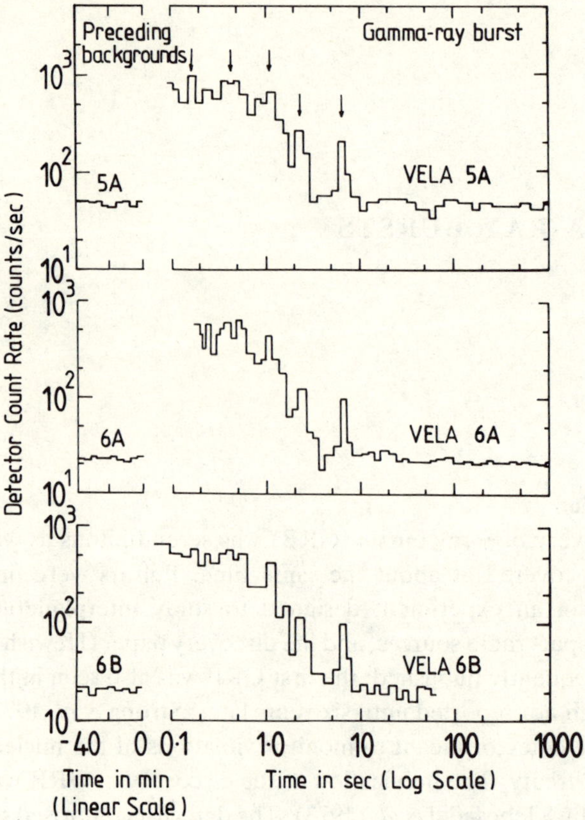

Figure 3.1. Count rate as a function of time for GRB 700822 as recorded at three VELA spacecraft by Klebesadel *et al*. (1973). Arrows indicate some of the common structure. Background rates immediately preceding the burst are also shown. VELA 5A count rates have been reduced by 100 counts s^{-1} (a major fraction of the background) to emphasise structure.

at least two different spacecraft heralded the discovery of the phenomenon of GRB.

In their discovery paper, Klebesadel *et al*. (1973) reported the detection of 16 bursts during the three-year period July 1969 to July 1972. The time histories of one of the bursts (GRB 790822†) as seen by detectors on three separate spacecraft are shown in Figure 3.1. The near coincidence in time of occurrence of an order of magnitude increase in the counting rates and

† It is customary to denote a GRB by the notation GB or GRB ijklmn, where ij, kl and mn refer to the year, the month and the date on which the burst occurred. If more than one burst was seen on a given date, an additional symbol a, b, c . . . is added in the order (in time) in which they occurred.

similarities in burst profile features in more than one detector left no doubt about the reality of the phenomenon. The burst shown in Figure 3.1 had an integrated flux density $\sim 8 \times 10^{-5}$ erg cm^{-2}. The arrival times of the event at different spacecraft were recorded to an accuracy of 0.05 s. From the finite differences in these times, it was possible to deduce crudely the arrival direction of the GRB. From such knowledge, the Sun and Earth were ruled out as the sources of the 16 GRB reported, though an additional burst from the direction of the Sun was seen. Colgate (1968) had predicted, before the discovery was announced, that intense emission of prompt γ-rays could occur during the initial stages of the development of SN; however, there does not seem to be any correlation between the GRB and reported SN seen at optical wavelengths and another mechanism is necessary.

Following the announcement of the discovery of GRB, groups all over the world built dedicated instruments to study this exciting phenomenon and these have been flown in geocentric and heliocentric orbiting spacecraft and on planetary probes in order to have improved direction information.

To date more than 300 GRB are positively known and a few hundred more have probably also been detected, but we await their analysis and publication.

GRB are highly individualistic, each being unlike any other in some features. Their time histories are much more disparate than their energy spectra, which can in themselves be quite distinctive. Any gross description of GRB as a class has therefore to be necessarily crude. One such description follows: the size or 'fluence' (integral of flux over time) for most of the GRB is in the range 10^{-7}–10^{-3} erg cm^{-2}. At a typical flux of 10^{-5} erg cm^{-2} s^{-1}, and a typical distance of 1 kpc, the luminosity ($\sim 10^{39}$ erg s^{-1}) surpasses for a brief time that of the diffuse background γ-radiation over the entire Galaxy. The duration of a GRB ranges from a few tens of milliseconds to a few hundreds of seconds. Except in two cases (GRB 790305 and GRB 790324) there is no direct and strong evidence to show that they recur. Some of the bursts exhibit absorption (intensity lower than the continuum) or emission (intensity higher than the continuum) features in their energy spectra.

The single most remarkable GRB (GRB 790305) occurred on March 5, 1979 (Mazets *et al.* 1979*a*, *b*). The question whether it is truly unique or an extreme example of a class of bursts is not yet settled. This GRB merits a mention of a few of its features. It is the most intense GRB observed to date at a fluence of $>10^{-3}$ erg cm^{-2} and has the smallest risetime. An 8 s periodicity in emission following the impulsive phase was established by several groups and a quasi-periodicity with a period 23 ms has been claimed by one group (Barat *et al.* 1983). It was seen by no fewer than nine different detectors. Its source direction was the most precisely determined among all the GRB, with an error box ~0.1 arc min^2 (Cline *et al.* 1982). Apparently it is associated with the SNR N49 in the Large Magellanic Cloud situated at a distance of 55 kpc (see the reviews by Cline 1980, 1982). The Leningrad group, however, has persistently questioned this association (see, for example, Mazets and Golenetskii 1981*a* and Golenetskii *et al.* 1984), arguing that, at source distances greater than 1 kpc, the Eddington luminosity limit is violated. If the association is true, it makes this object the only extragalactic GRB source detected so far. This GRB source is also one of the two clear cut cases of recurrent bursts.

Although there was initially a plethora of potential models for GRB sources (see, for example, Ruderman 1975), the majority opinion today favours highly magnetised neutron stars as the GRB sources. The actual mechanism of γ-ray production is not yet agreed upon, but the basic energy source powering the burst is expected to be gravitational or nuclear binding energy.

References to much of the experimental and theoretical work on the subject can be found in papers published in the proceedings of the following conferences:

(*a*) the 1979 Toulouse conference (Fance), *Astrophys & Space Sci.* (1981) **75**, 5–224; (*b*) the 1980 Royal Society Meeting (UK), *Phil. Trans. Roy. Soc. Lond.* (1981), **A301**; (*c*) the 1981 La Jolla Conference (USA), *Am. Inst. Phys. Conf. Proc. 77* (1982) (Lingenfelter, R. E., Hudson, H. S., and Worrall, D. M. eds); (*d*) the 1982 Garching Conference (FRG), *Max Planck Institute Report 177* (1982) (Brinkmann, W., and Trumper, J. eds); (*e*) the 1983 Goddard Conference (USA), *Am. Inst. Phys. Conf. Proc. 101* (1983) (Burns, M. L., Harding, A. K., and Ramaty, R. eds); and (*f*) the 1984 Santacruz Conference (USA), *Am. Inst. Phys. Conf. Proc. 115* (1984) (Woosely, S. E. ed).

The subject of GRB is one of the areas in astrophysics that is being pursued most vigorously both experimentally and theoretically. In the next section we will present a few details of the detectors that are employed in their study. Typical results on time histories and energy spectra are presented in Sections 3 and 4, respectively. Section 5 deals with efforts to identify GRB source objects with those seen at other wavelengths. This is followed by a presentation of the statistical distributions of GRB (in intensity and in galactic coordinates) in Section 6. We will then present a discussion of source models in Section 7 before ending the chapter with a brief summary.

3.2 Detectors and instrumentation

The guiding principles in the design of the recent GRB detectors have been (i) to attain as low a detection threshold as possible, affording nearly bias-free statistical analysis of GRB; (ii) to have a good energy resolution, allowing better delineation of spectral features; (iii) to have accurate event time information, resulting in a precise source direction determination; and (iv) to record many energy spectra over as short an integration time as possible to study fast spectral variability. Some of the detectors are mentioned below together with references where information can be found on them:

> KONUS experiments of the Leningrad group (Mazets and Golenitskii 1981*b*); SIGNE experiments of the Franco–Soviet collaboration (Barat *et al.* 1981); ISEE-3, the International Sun–Earth Explorer Experiment (Klebesadel *et al.* 1981 and Teegarden and Cline 1981); SMM, the Solar Maximum Mission Experiment (Forrest *et al.* 1980); HEAO-1 (Knight *et al.* 1981); PVO, the Pioneer Venus Orbiter Experiment (Evans *et al.* 1981); and a balloon-borne experiment (Fishman 1981) which is a prototype of BATSE (Burst and Transient Source Experiment) on the GRO scheduled for a 1988 launch. This list is admittedly incomplete but hopefully representative.

3.2.1 Detector description

While referring the reader to the list of references given above, we elaborate here on just two detectors chosen to illustrate the kind of instrumentation that is generally employed.

ISEE–3 Ge detector. The high resolution Ge detector (Teegarden and Cline 1981) is one of several sensors aboard the ISEE–3 spacecraft launched in September 1978 into a halo orbit about the Lagrangian point approximately 230 Earth radii radially inward towards the Sun. The high purity germanium crystal, 4.02 cm in diameter and 2.9 cm thick, is shielded from direct sunlight. It is also thermally insulated from the rest of the spacecraft and is radiatively cooled, so that it remains at a constant temperature of 130 K. It has a resolution ~10 keV at 570 keV. The useful energy window is from 200 keV to 3 MeV. Instead of storing the number of counts in a succession of fixed time intervals, elapsed times for the occurrence of a ground commandable fixed number of counts (time-to-spill mode) are recorded with a command-selectable resolution of 0.12 to 1.0 ms. If the time-to-spill does not exceed a ground-selectable value, a trigger is generated to initiate the data recording routines. Energy spectra are recorded with a time resolution of 8 ms. The 10^5 bit on-board memory, in which all this information is recorded, is dumped on to an earth station either automatically upon filling or, manually, on a ground command.

KONUS detectors. A block diagram of the sensors and instrumentation of the KONUS experiment (Mazets *et al*. 1983) aboard the spacecrafts Venera 13 and Venera 14 is shown in Figure 3.2. There are six detectors on each spacecraft facing the six directions of a Cartesian coordinate system. Each detector is a NaI(Tl) scintillator of diameter 8.0 cm and 3.0 cm thick and each has a passive shield on the sides made up of lead and tin. Background rates from each of the six sensors are monitored once in 20 min in the 45 to 200 keV energy range. When the counting rate over an integration time of 0.25 s significantly exceeds the average rate a trigger is generated to initiate the processing of information from the six detectors. The time analyser stores the 8 s long prehistory with a resolution of 0.25 s and measures the burst time profile for 0.875 s with a

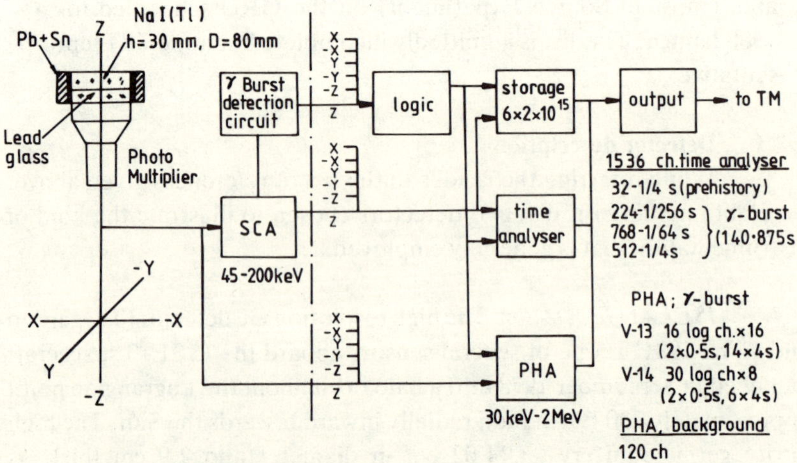

Figure 3.2. Block diagram of KONUS instrumentation for detectors aboard the spacecraft Venera 13 and 14. (Mazets *et al*. 1983.)

resolution of 1/256 s, then for the next 12 s with a resolution of 1/64 s and finally for a further 128 s with a resolution of 1/4 s. Thus 8 s of prehistory and a total 140.875 s of post-trigger time profile of each GRB is recorded with a varying time resolution. γ-ray counting rates from each of the six detectors during the burst are recorded in 0.5 and 8 s time intervals. This information, when combined with the measured angle-dependent efficiency of each detector (which approximates to a $\cos\theta$ distribution, where θ is the angle of incidence of a GRB with respect to the normal to the detector face), yields information on the source direction. In order to have good statistics, and yet maintain correct energy assignment, the pulse height analysers (PHA) are connected to the single detector that is most favourably oriented with respect to the burst source. The PHA on Venera 13 records 16 energy spectra over as many consecutive time intervals, two of 0.5 s and 14 of 4 s duration; each spectrum has 16 quasi-logarithmic energy bins. The 30 channel PHA on Venera 14 records eight consecutive energy spectra, two of 0.5 s and six of 4 s duration.

3.2.2 Selection effects and observational biases

Spurious events. First, there is the important problem of spurious events seen in only one detector that have nothing to do with the phenomenon of GRB. For example, Mazets and Golenetskii (1981*a*) report several spurious bursts falling in this category. These are thought to be due to phosphorescence in the NaI(Tl) crystal induced by high energy losses that occur in the detector when multiply charged cosmic rays pass through it or when nuclear spallation takes place (Fishman *et al.* 1978).

In an experiment with two simultaneous balloon flights separated by several hundred kilometres, Cline *et al.* (1977) observed the occurrence of a wide variety of fluctuations in detectors on one balloon but not on the other simultaneously. These variations are presumably of magnetospheric origin and underline the need for a GRB to be seen by two or more detectors located far apart for it to be accepted as such.

Energy spectra. In any given experiment one actually records an energy loss spectrum, and this is related to the incident photon spectrum by the relation

$$N(E') = \int_{E'}^{\infty} F(E)P(E, E')\,\mathrm{d}E \tag{3.1}$$

where E' is the energy loss detected; E is the energy of the γ-ray; and $P(E, E')$ is the probability that a photon of energy E gives rise to an energy loss E' in the detector. The observed energy loss spectrum has to be de-convoluted to obtain the true photon spectrum on the basis of laboratory calibrations of the $P(E, E')$ matrix and a trial (assumed) photon spectrum, $F(E)$, for the GRB. It is of course the de-convolved photon spectrum that forms the basis of any modelling.

Selection effects. Most of the experiments operate in the trigger mode; i.e. when the number of counts from a detector over a fixed integration time, τ, exceeds the background counts, $n_b\tau$, by a pre-determined number, k, of standard deviations, a trigger is generated to initiate data recording. If the burst duration, T, is less than or equal to τ, then the detection threshold is given by

$$S_0 = k(n_b\tau)^{1/2} \tag{3.2}$$

If, on the other hand, the burst duration is much longer than τ (with no spikes in the burst) then the detection threshold becomes

$$S_T = S_0 T/\tau \tag{3.3}$$

i.e. it increases linearly with burst duration time T. Thus, the detection threshold of a given experiment is a function of τ and k (which are experimenter settable) and T and this introduces a selection effect. A moderately intense burst lasting for a very long time may never be recorded in an experiment because of this effect. Mazets and Golenetskii (1981*a*) have treated this problem mathematically and found that, for a flat burst–duration distribution $\phi_T = 1/T_{max}$ for $0 \leqslant T \leqslant T_{max}$, the efficiency of recording a burst with a fluence, S_0, is given by

$$\varepsilon(S_0) = \frac{\tau}{T_{max}}\left[3 - 2\,(\tau/T_{max})^{1/2}\right] \tag{3.4}$$

for a burst size distribution of the type $N(>S) \propto S^{-3/2}$ (see Section 3.6). For an S^{-1} distribution, the efficiency is given by

$$\varepsilon(S_0) = \frac{\tau}{T_{max}}\left[1 + \ln\,(T_{max}/\tau)\right] \tag{3.5}$$

Here $\varepsilon(S_0) = N_{obser}(S_0)/N_{true}(S_0)$. For example, if $\tau = 0.25$ s and $T = 3$ s, the detection efficiency for a burst with a fluence S_0 is 0.20 for $S^{-3/2}$ distribution and 0.29 for S^{-1} distribution. For very intense bursts, $S_0 \leqslant S \leqslant S_{max}$, the detection efficiencies are likewise given by

$$\varepsilon(S) = \frac{S}{S_{max}}[3 - 2\,(S/S_{max})^{1/2}] \tag{3.6}$$

for an $S^{-3/2}$ distribution, and by

$$\varepsilon(S) = \frac{S}{S_{max}}[1 + \ln\,(S_{max}/S)] \tag{3.7}$$

for an S^{-1} distribution. While comparing data from different experiments, it is necessary to keep the above selection effects in mind. These effects have obviously a role to play in the interpretation of log N – log S curves (see Section 3.6).

3.3 Time histories

Time profiles provide important clues to understanding the nature of the GRB sources. As an illustration, it is well known that GRB 790305 exhibits an 8 s periodicity (Mazets *et al.* 1979*a*, *b*; Barat *et al.* 1979; Cline *et al.* 1980; and Terrell *et al.* 1980). If this periodicity is due to the rotation of the burst source, one expects the centrifugal force at the surface of the source to be less than the gravitational force, which leads to a matter density

$$\rho \geqslant 3\pi/GP^2 = 2.2 \times 10^6 \text{ g cm}^{-3} \tag{3.8}$$

where P is the period of rotation of the source object, and G is the gravitational constant. Densities in this range imply that the object has to be a white dwarf, a neutron star or an even more compact object. Furthermore, the very short risetime, observed for this GRB ($\leqslant$1 ms), implies an emission region with linear dimensions $\leqslant$1000 km, which, when combined with a canonical 1 solar mass, leads to densities $>10^9$ g cm^{-3}, leaving only neutron stars or higher density gravitationally collapsed objects as candidate sources for the GRB. Fast time scales also impose additional constraints on models in that they should adequately describe correspondingly fast energy transport to the emission regions. The implications of time histories for GRB source models are postponed to Section 3.7.

3.3.1 Burst profiles and durations

No two GRB are completely alike in their time profiles. To illustrate we present in Figure 3.3 profiles of three GRB with three different burst durations (Hurley 1983*a*). In Fig. 3.4 is shown the profile of a very short duration burst (Barat *et al.* 1984*a*). The time profile of

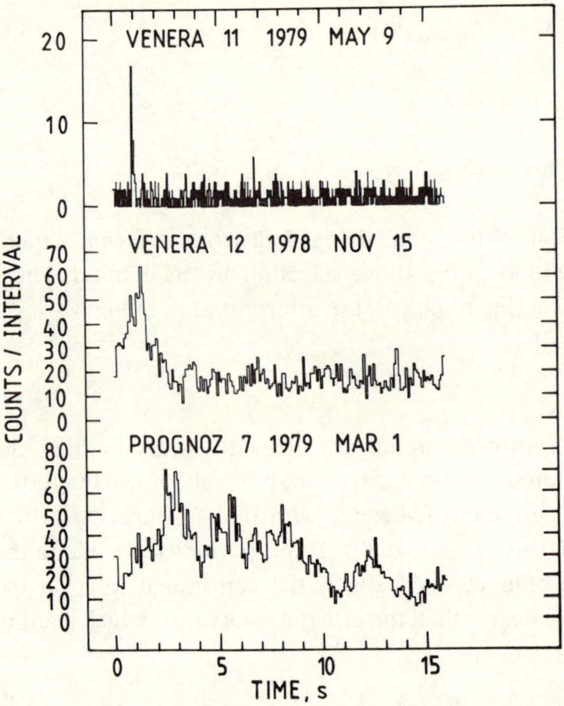

Figure 3.3. Three GRB histories from the Franco–Soviet SIGNE experiments, illustrating the three classes of bursts: very short (100 ms) events, long (1 s) events, and long complex events. The time histories are for the energy range 100–600 keV (Hurley 1983*a*).

GRB 790305, shown in Figure 3.5, is unique. Note the very fast risetime ($\leqslant$0.25 ms), the initial intense peak lasting for about 130 ms and a series of pulsations following the impulsive peak. The peak intensity of this burst, at several times 10^{-3} erg cm^{-2} s^{-1}, is the highest recorded yet.

A remarkable 20-minute-long γ-ray transient (the so-called Jacobson transient) was observed by Ling *et al.* (1982) in a balloon-borne detector on June 10, 1974. Four intense γ-ray lines at 0.413, 1.79, 2.2 and 5.9 MeV appeared in the background spectrum. The widths were all broader than the instrument resolution. No continuum emission was detected. Lingenfelter *et al.* (1978) suggested that the 0.413, 1.79 and 5.9 MeV lines are the red-shifted lines (z = 0.20–0.29) due to pair annihilation and neutron capture on hydrogen and iron, respectively, at and near the surface of neutron star accreting matter in a binary system. The 2.2 MeV line, on the other hand, was interpreted as the unshifted line from neutron capture on

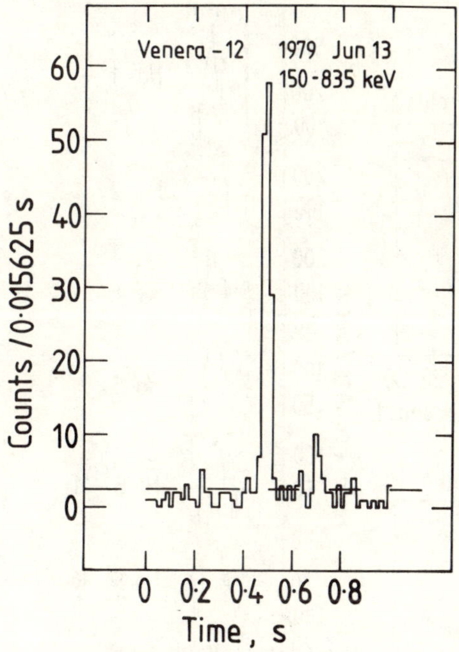

Figure 3.4. Time histories of GRB 790613 from the Franco–Soviet SIGNE experiments for the energy range 150–835 keV (Barat *et al.* 1984*a*).

hydrogen in the atmosphere of the companion star. Subsequent searches failed to observe similar events. It is not clear what relationship, if any, exists between GRB and the Jacobson transient.

Cline and Desai (1974) have pointed out that very brief, ~0.1 s, bursts form a class distinct from the long and highly structured bursts. Mazets *et al*. (1982*b*) suggest that there is a distinct class of short duration bursts of which GRB 790305 is the prime example. In Figure 3.6, we have shown the distribution of burst durations for the 143 GRB observed by the Leningrad group (Mazets and Golenetskii 1981*a*) with detectors aboard Venera 11 and 12. There appears to be a distinct break in the distribution at 1 s. More recent results from the same group, as quoted by Norris *et al.* (1984), from detectors on Venera 13 and 14 seem to confirm this. Based on results from ISEE–3, Norris *et al*. (1984) present additional evidence for a break at 1 s in the burst duration distribution. Barat *et al.* (1984*a*) point out that there is a break at 600 ms in the burst duration distribution of single peaked bursts. Norris *et al*. (1984), emphasising the importance

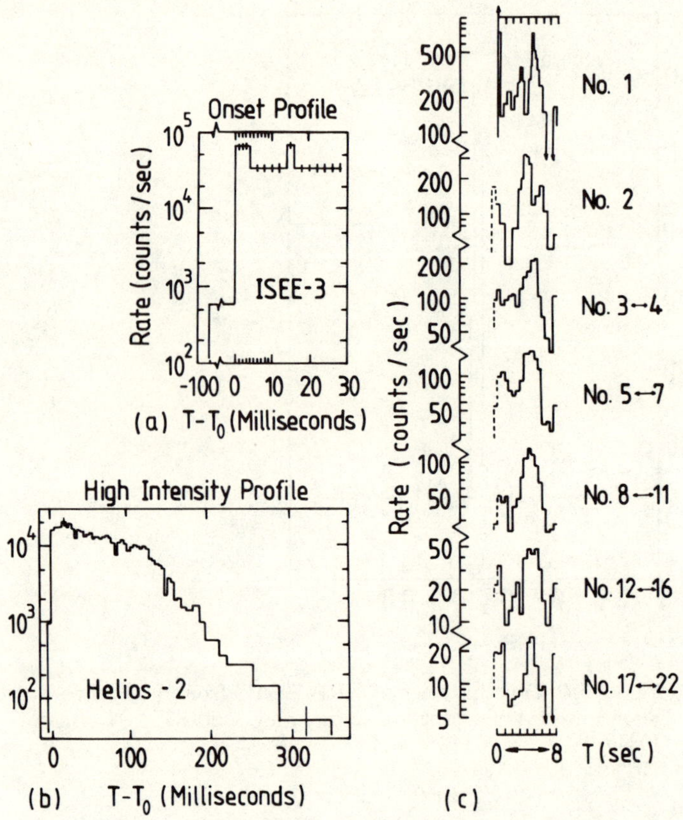

Figure 3.5. Time histories of GRB 790305. (*a*) The onset of the high intensity portion of the burst showing the increase in count rate by two orders of magnitude within a resolving time of 1 ms (ISEE-3). (*b*) Details of the high intensity portion of the burst observed with HELIOS-2. (*c*) The first 22 cycles of the burst plotted on an 8 s period basis with the event onset chosen as zero of time folded with increasing number of cycles per plot. The initial peak remains at constant phase within 1 s yielding an average period of 8.0 ± 0.05 s (Cline *et al.* 1981*a*).

of selection effects in recording short duration bursts, give figures ranging from 0.07 ± 0.02 to 0.29 ± 0.10 for the fraction of GRB with $T \leqslant 1$ s among all the bursts. If this classification is true, it implies that distinct processes or source populations are responsible for the long and short bursts.

Mazets *et al.* (1982*b*) point out that the short duration bursts have many properties in common; in particular, they point out that their energy spectra are all soft with $kT \sim 35$ keV. There are, however, observations

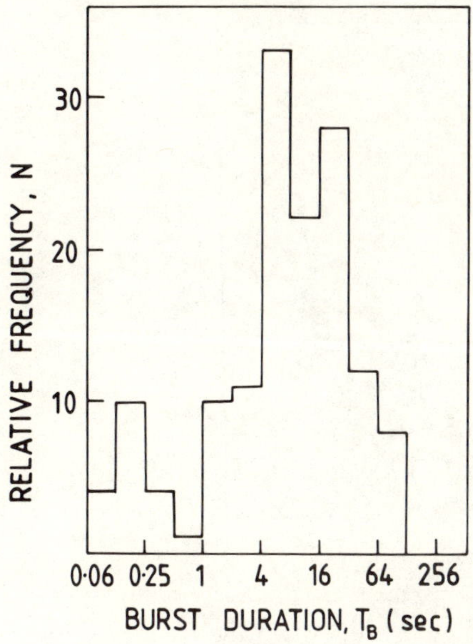

Figure 3.6. Distribution of 143 GRB with respect to duration as obtained in the KONUS experiments on Venera 11 and 12. Ordinate: number of bursts per interval $\Delta \log T_B = 0.3$ (Mazets and Golenetskii 1981*a*).

contradicting this; see Laros *et al.* (1981, 1982) and Barat *et al.* (1984*b*). Furthermore, Mazets and Golenetskii (1981*a*) find that a positive correlation exists between the burst fluence and its duration; see Figure 3.7. This observation is corroborated by Barat *et al.* (1984*a*).

3.3.2 Burst risetimes

The well known GRB 790305 has the shortest detected risetime, $T_r < 0.25$ ms (Cline *et al.* 1981*a*). Barat *et al.* (1984*a*) have studied the distributions of risetimes, T_r, decay times, T_d, and peak fluxes of a selected sample of 20 short single peaked GRB. These authors suggest that the rise- and decay times, being less subject to instrumental effects, are better parameters to describe a GRB than the burst duration, T. They find a correlation between T_r and T_d among the 20 GRB, as shown in Figure 3.8. The distributions, both T_r and T_d, seem to be continuous, extending over several orders of magnitude, and suggest that there is nothing like a standard burst profile. A wide variety of physical conditions may therefore be present at or about the burst sites. It is difficult to know if time histories directly reflect processes at work at the site of energy

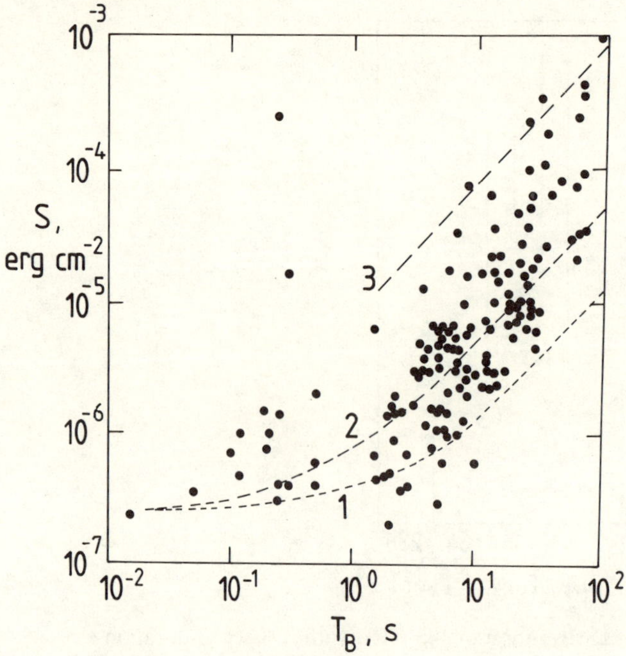

Figure 3.7. Intensity of GRB as a function of their duration, as seen in the KONUS experiments. Curve 1 shows the sensitivity limit; curve 2 is the calculated instrumental sensitivity; and curve 3 is the boundary of the distribution defined by the sources nearest the Sun (Mazets and Golenetskii 1981*a*).

release or whether, for example due to beaming, they carry a signature of conditions at a more distant site. Barat *et al.* (1984*a*) found that T_r and T_d when expressed in units of milliseconds are related by

$$T_r = 0.13T_d^{1.25} \tag{3.9}$$

at $T_r > 70$ ms. In general $T_r \sim T_d$ within a factor of two. One notable exception is the GRB 790305 for which the risetime is two orders of magnitude smaller than that of both an average short duration burst and its own decay time. Barat *et al.* found that the mean risetimes are 17 ± 3 ms and 402 ± 40 ms and the mean decay times are 19 ± 5 ms and 568 ± 56 ms for short ($\sim$100 ms) and long ($\sim$1 s) duration bursts, respectively. They also found that the peak flux during a burst is a decreasing function of risetime; see Figure 3.9. The authors noted that bursts with the following combinations of T_r and T_d (which *could* have been recorded) do not occur in nature: (i) short T_r, long T_d; (ii) long T_r,

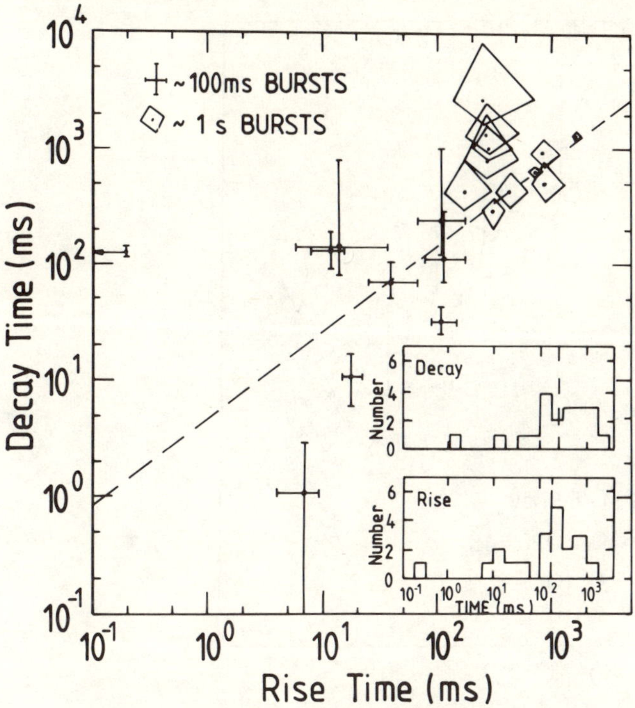

Figure 3.8. e-folding rise and decay times of the 20 single peaked GRB studied by Baret *et al.* (1984*a*). Error bars are for 68 % confidence level. Crosses are for shorter duration (~100 ms) bursts, and diamonds are for the longer bursts. Dashed line: (risetime) = 0.13 (fall time)$^{1.25}$. Insets are histograms of the numbers of events as a function of rise and decay times.

short T_d; and (iii) long duration bursts with short T_r and short T_d. One is cautioned, however, that the sample is small, limited to only 20 events and instrumental effects may also play a role.

3.3.3 Burst periodicities and quasi-periodicities

Periodicity in the emission of γ-rays can arise due to rotation or precession of the burst source or due to plasma oscillations surrounding it. As will be clear from the following, periodicity in a GRB is the exception rather than the rule.

The most convincing example of a periodicity in GRB emission is the well known GRB 790305. Mazets *et al*. (1979*a*, *b*) were the first to report

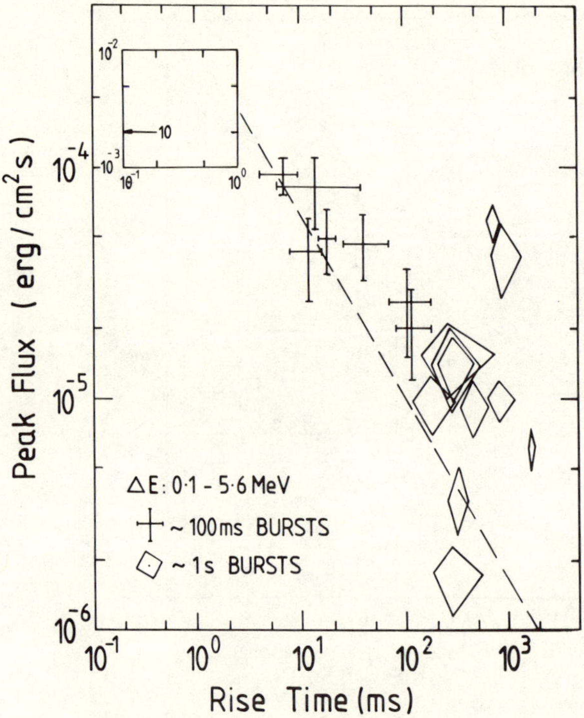

Figure 3.9. The peak fluxes are shown as a function of risetime for 19 of the single peaked GRB studied by Barat *et al.* (1984*a*). Dashed line: (peak flux) = 4×10^{-4} (risetime)$^{-0.8}$. The peak fluxes have been extrapolated to a common energy range 0.1–5.6 MeV.

a periodicity, the period being 8.1 ± 0.1 s; see Figure 3.10. This result was soon confirmed by Barat *et al.* (1979) with $P = 7.9 \pm 0.3$ s, by Cline *et al.* (1980) with $P = 8.0 \pm 0.05$ s, and by Terrell *et al.* (1980) with $P = 8.0 \pm 0.05$ s. Cline *et al.* detected pulsations out to 172 s after the initial impulsive peak; see Figure 3.5*c*. The initial impulsive peak occurred at a phase coincident with that of the interpulse at a phase of 0.5 (Mazets and Golenetskii 1981*a*). The latter authors found that pulsations ceased sometime prior to 360 s after the commencement of the burst. The exact point in time when an object ceases pulsating is doubtless governed by processes near the object itself. Its measurement, however, is dependent on the geometrical acceptance factor of the sensor in a given experiment, i.e. on its ability to distinguish the signal from fluctuations in the background noise and also on the capacity of the memory bank aboard the spacecraft to hold a long and continuous data stream. Mazets *et al.*

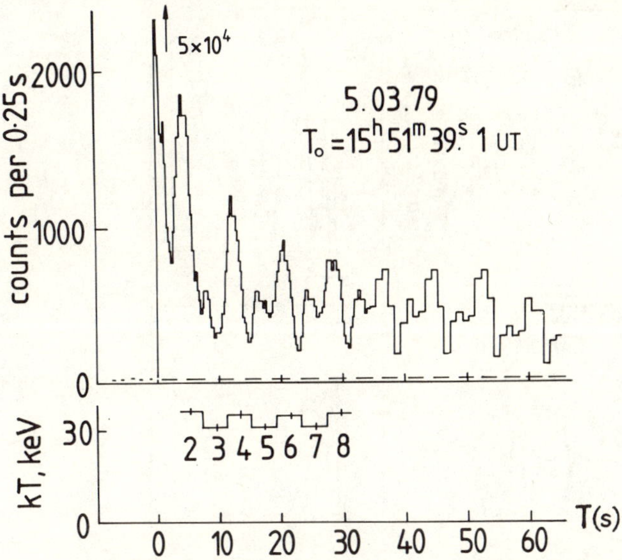

Figure 3.10. The pulsating phase of the time history profile of GRB 790305 as seen by the Leningrad group in the KONUS experiments. Top: time profile showing narrow initial pulse and pulsations with a resolution of 0.25 s. If the main pulse in the pulsations (i.e. the second big peak) is assigned zero phase, the narrow initial pulse has a phase 0.5. Bottom: temperature variations for the thermal bremsstrahlung spectra (integrated over 4 s) in the pulsating stage of the burst, $I \propto \exp(-E/kT)$ (from Mazets *et al.* 1982*b*).

(1981*a*) found that during the first four periods the main pulse weakens in intensity while the interpulse stays at the same level. During the next three pulsations the situation is reversed.

The second convincing example of a periodicity is the GRB 771029 for which Wood *et al.* (1981) found a period of 4.2 ± 0.2 s. The data exhibit the periodicity over five to six cycles.

In addition, a number of other GRB are also reported to exhibit periodicities or quasi-periodicities. These are GRB 781104 ($P = 0.280$ s) and GRB 781023 ($P = 0.143$ s) by Chambon (1982); GRB 771029 ($P = 4$ s) by Pizzichini (1981); GRB 790113 ($P = 5.7$ s) by Barat *et al.* (1984*d*); GRB 781102 ($P \sim 15$ s) by Mazets *et al.* (1982*b*); and a few other cases by Loznikov and Kuznetsov (1982). Not all cases of reported periodicity are convincing, however. One must recognise here the experimental difficulty in demonstrating the existence of a periodicity when the length of the data stream is not much longer than the suspected period

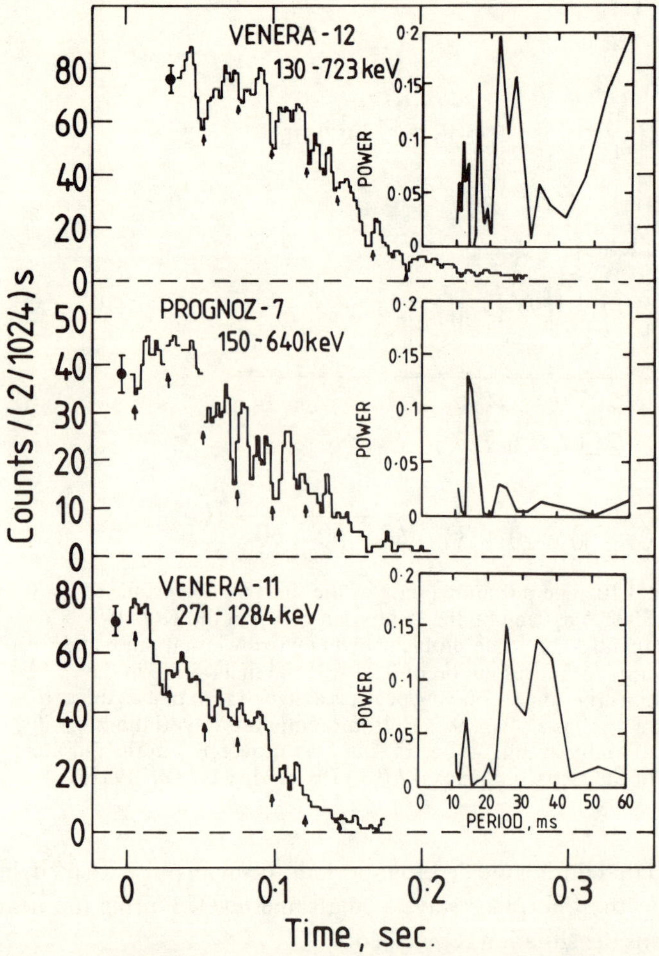

Figure 3.11. Time profiles of GRB 790305 during the first 200 ms observed with 2/1024 s time resolution by identical detectors aboard the Prognoz 7, Venera 11 and Venera 12 spacecraft (Barat *et al*. 1983). Dashed lines: background levels; arrows: 23 ms period. Insets: spectral power in counts^2 Hz^{-1} as a function of period. These spectra result from a power spectral analysis of the residuals of the raw data.

itself and there is but a single opportunity to observe the burst since the phenomenon by its very nature is transient.

Barat *et al.* (1983) suggest that GRB 790305 may also be exhibiting a periodicity of 23 ms (Figure 3.11) on the basis of power spectral and superposed epoch analyses of their data on the time history of the burst. Following Brecher (1982), one can think of 23 ms as the rotation period

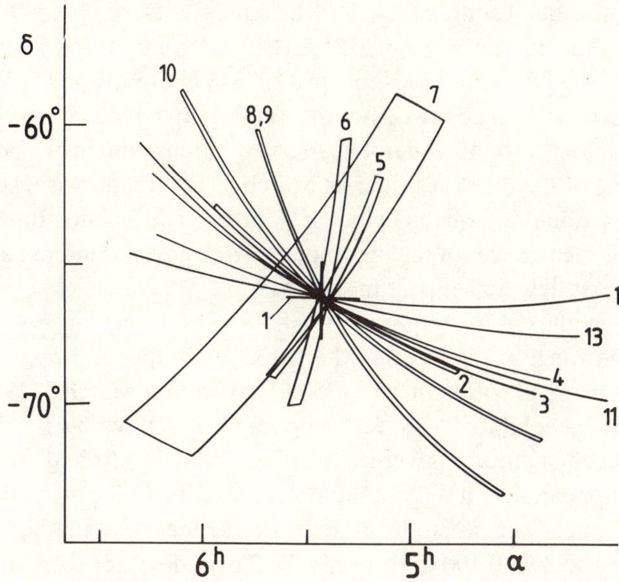

Figure 3.12. GRB recurrences of the source of the famous GRB 790305, as seen by the Leningrad group (Golenetskii *et al.* 1984). The localisation error band is shown for each recurrence. Numbers 1 to 13 on the bands refer to the events on March 5, March 6, April 4, April 24, 1979, December 1, 1981, January 2, February 15, April 23 *a* and *b*, May 10, May 30, July 8 and December 31, 1982, respectively.

of the neutron star (the source object of GRB 790305) and 8 s its precession period.

3.3.4 Burst recurrences

Since more and more GRB are being detected as time progresses, the problem of chance superposition of source directions within the error boxes is getting worse. It is therefore imperative to determine burst source directions rather precisely to be able to say unequivocally if a particular source is recurrent or not. To date there are only two sources known to be recurrent: GRB 790305 and GRB 790324.

An examination of Venera 11 and 12 data by Mazets *et al.* (1979*a, b*) suggested that GRB 790305 showed itself again on March 6, April 4, and April 24, 1979. More recently the same group, Golenetskii *et al.* (1984), have presented evidence for 12 more recurrences from this source based on Venera 13 and 14 data. Intersection of the error bands for the source direction in each of the 16 events at the same position, as shown in Figure 3.12, leaves no doubt about the identity of the source. The published recurrences of this burst source (RA = 05 h 25 m 55 s, Decl. = $-66°07'$) occurred during the time spans from March 1979 to January 1980 (Venera 11 and 12) and from December 1981 to April 1983 (Venera 13 and 14) with a break in observations of about 22 months. If the source recurred during this break, it almost certainly did not have a very intense impulsive peak at the beginning of the burst as it had on March 5, 1979; otherwise, such an intense burst would have been detected by other experiments. Indeed the 15 recorded recurrences of the burst source were never as intense as it was at its very first detected appearance on March 5, 1979.

The fluence, in units of 10^{-6} erg cm^{-2}, was 450 at its first appearance and generally in the range 0.15–7 during the subsequent appearances. The energy spectra were all soft with $kT \sim 35$ keV except on March 5, 1979, when the spectrum exhibited a harder component as well, with $kT = 520 \pm 100$ keV. The burst durations were in the range 100 ms–3.5 s and except during its first appearance, it did not show any pulsations. The mean time between recurrences appears to be 38.6 d with the least and the highest intervals being 12.8 h and 100 d, respectively. There does not seem to be any correlation between the time interval between two consecutive recurrences and the corresponding fluence values.

In a recent paper Rothschild and Lingenfelter (1984) have noted a pattern in the recurrence times of these bursts, showing an apparent period of 164 d. The pattern is that their recurrence time (time elapsed since the previous burst) keeps progressively increasing and then becomes very small, giving a saw-tooth curve with a period of 164 d. The authors suggest that the pattern reflects periodic accretion on to a neutron star from a binary companion in an eccentric orbit ($e \approx 0.9$). Rothschild and Lingenfelter propose that this system drives two types of GRB: (i) the relatively frequent but weak recurrent bursts produced by either magnetospheric gating, or thermonuclear runaway, of periodically accreting matter around the neutron star; and (ii) the much rarer intense bursts, such as that of March 5, 1979, produced by tidally triggered star-quakes. If this pattern is confirmed, it represents perhaps the first evidence for a binary companion around a burst source.

The only other example of a recurrent source is the object of GRB 790324 (Golenetskii *et al.* 1979; Mazets *et al.* 1979*a*,*b*; and Mazets *et al.* 1981*a*). This source recurred on March 25, and March 27, 1979. The fluences on the three occasions, in units of 10^{-6} erg cm^{-2}, were 1, 1.5 and 0.3, respectively.

Among the 40 burst sources precisely localised by the International Network (HELIOS-2, SIGNE, PVO and ISEE-3) no recurrent bursts were observed over a period of ~1.5 y (Vedrenne 1983). Taking the solid angle of coverage into account, Vedrenne states that the recurrence time for a GRB is more than one year. The well known GRB 790305 which recurs even on time scales of less than a day is, of course, an exception.

Terrell *et al.* (1980) are of the view that the recurrences of GRB sources are related to the characteristics of the neutron star itself such as an internal restructuring or a critical accumulation of accreted material and not due to the improbable recurrent collisions with incoming solid objects such as comets or asteroids.

3.4 Energy spectra

3.4.1 Spectral measurements and fits

Early results on the energy spectra of GRB were given by Cline *et al.* (1973) based on their IMP-6 data. These spectra could all be fitted with a simple exponential of the type

$$\mathrm{d}N/\mathrm{d}E = I_0 \exp(-E/E_0) \text{ photons cm}^{-2}\, s^{-1}\, \text{keV}^{-1} \tag{3.10}$$

where $E_0 \sim 150$ keV. Later, Cline and Desai (1975) represented the data from IMP-7 by the original 150 keV exponential modified by a power law of the type $\sim E^{-2.5}$ fitted to the high energy tail of the burst. Gilman *et al.* (1980) fitted the spectrum of the Apollo 16 event GRB 720427 (Metzger *et al.* 1974) with an optically thin thermal bremsstrahlung spectrum of the type

$$\mathrm{d}N/\mathrm{d}E = AE^{-1}g(E) \exp(-E/kT) \text{ photons cm}^{-2}\, s^{-1}\, \text{keV}^{-1} \tag{3.11}$$

where g is the energy dependent Gaunt factor of the order of unity; T is the temperature of the plasma; and k is the Boltzmann constant with $kT \approx 500$ keV. Using their most comprehensive data base (Mazets *et al.* 1981*b*) from the KONUS experiments, Mazets *et al.* (1981*c*, 1981*d*) adopted thermal bremsstrahlung-like spectra with kT values in the range 10 keV–2 MeV, most of them being 150–400 keV. A few examples of these thermal bremsstrahlung fits are shown in Figure 3.13. The Solar Maximum Mission group (Rieger *et al.* 1982) found a few cases of GRB in which photons with energies greater than 10 MeV were seen. The spectral fit for one of them, GRB 820330, is shown in Figure 3.14; it is given by a power law spectrum of the type

$$\mathrm{d}N/\mathrm{d}E = AE^{-2.5} \text{ photons cm}^{-2}\, s^{-1}\, \text{keV}^{-1} \tag{3.12}$$

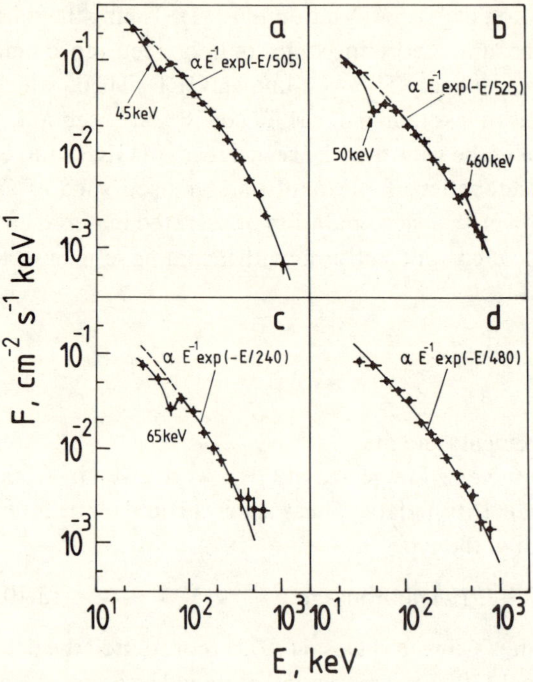

Figure 3.13. Typical γ-ray lines recorded in the energy spectra in the KONUS experiments (Mazets *et al.* 1980, Mazets and Golenetskii 1981*a*). (*a*) GRB 790307, (*b*) GRB 790622, (*c*) and (*d*) GRB 791101 spectra obtained during the first 8 s and the subsequent 24 s, respectively, demonstrate how the cyclotron absorption line evolves with time.

in the energy range 0.3–15 MeV. Notice that the energy range here does not overlap those in most other experiments.

Liang *et al.* (1983) have argued that in the physical conditions obtained in the emission regions, thermal synchrotron spectra resulting from mildly relativistic electrons in an optically thick plasma (Petrosian 1981) associated with strong magnetic fields are more appropriate. They have shown several examples of excellent fits to the KONUS data based on thermal synchrotron emission. These are of the type

$$dN/dE \propto \exp\left[-(4.5E/E_c)^{1/3}\right] \qquad (3.13)$$

where $E_c = heB\tilde{T}^2/m_ec$; $\tilde{T} = kT_e/m_ec^2$; T_e is the temperature of the plasma; B is the magnetic field; E is the photon energy; and e and m_e are the charge and mass of the electron, respectively.

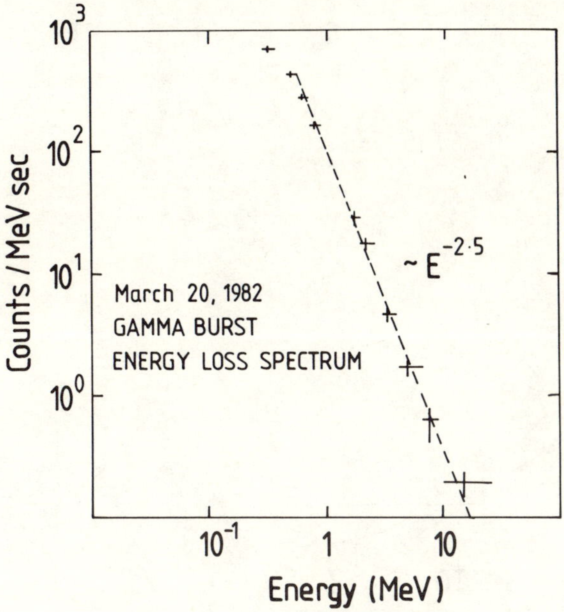

Figure 3.14. Energy spectrum of γ-rays from GRB 820320 as recorded by a CsI crystal in the SMM experiment (Rieger *et al*. 1982). The data can be represented by a power-law spectrum of the type $N(E) \propto E^{-2.5}$.

Fenimore *et al*. (1982*a*) argue that at temperatures ($kT \sim 300$ keV) implied by the bremsstrahlung fits, the free–free cross section is so much smaller than the inverse Compton cross section that the latter might dominate the spectral formation processes. They accordingly view GRB sources as Comptonised X-ray sources, i.e. sources in which an initial X-ray black body spectrum ($kT_{BB} \sim 2.4$ keV) has undergone inverse Compton scattering by a much hotter ($kT_e \sim 150$ keV) plasma. Although the authors offer no explanation as to why the hot plasma should be overlying an X-ray source, their Comptonised X-ray black body fits for the GRB 781104 (Fenimore *et al*. 1982*b*) are as good as or slightly better than the thermal bremsstrahlung fits.

From what has been mentioned above, one might form the impression that different mechanisms operate for different GRB leading to a multitude of possible spectral fits. It may well be so; the fits presented, however, do not prove it. This point is illustrated by Figure 3.15 from Liang *et al*. (1983) where equally good fits are obtained for GRB 780918, based on thermal bremsstrahlung, thermal synchrotron and inverse

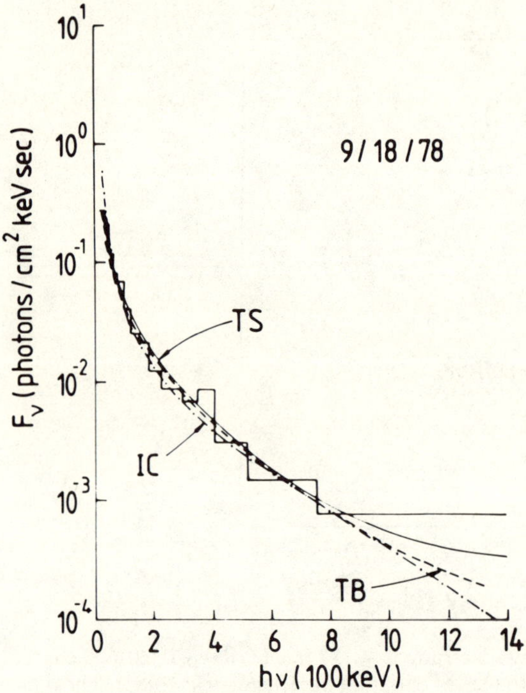

Figure 3.15. Good fits can be obtained by thermal bremsstrahlung (TB), thermal synchrotron (TS) and inverse Compton mechanisms (IC) to the same experimental data on energy spectrum of γ-rays from GRB 780918 (Liang *et al.* 1983).

Compton models. Furthermore, Barat *et al.* (1984*a*) have also shown that for two-thirds of their sample of 20 short, single-peaked GRB, all four spectral types (thermal bremsstrahlung, thermal synchrotron, exponential and power law) give acceptable fits. The point is that, given the freedom of parameters and poor statistics at the upper end of the spectrum, many laws fit the data over limited energy spans (in most cases $1\frac{1}{2}$ decades), leading to the impossibility of demonstrating the uniqueness of a fit by a given law.

Attention must also be drawn here to two other aspects. First, as shown in Section 3.4.3, the energy spectra sometimes evolve with time on scales as short as 30 ms. Faced with such fast changing spectra, one questions the validity of using the event-averaged spectrum as a basis to understand the physical processes operating in and around the burst sources; see, for example, Lamb (1982), Teegarden (1982) and Vedrenne (1984). Second, different groups observing the same GRB sometimes disagree over the shapes and absolute values of energy spectra. For example, SIGNE, KONUS and ISEE-3 disagree on the spectrum from GRB 781104 (Vedrenne and Chambon 1983); SMM and ISEE-3 on the spectrum from GRB 800419 (Vedrenne and Chambon 1983); and KONUS and PVO on the spectrum from GRB 790613 (Barat *et al.* 1984*b*).

It is interesting to know if γ-rays of energies $\gtrsim 1$ GeV are present in GRB. Bhat *et al.* (1982) carried out a ground-based experiment at Ootacamund that depended on the bunching of GeV γ-rays in micro-bursts (10–100 μs in duration) for their detection. During a 1.5 y period of its operation, five GRB that were seen by satellite experiments were in the view of the experiment but no GeV γ-rays were seen from any of them. From this, the authors have concluded that either (i) there are no γ-rays at GeV energies in the GRB in question, or (ii) if the GeV γ-rays are indeed present, the micropulsations needed for their detection are absent, or (iii) if the GeV γ-rays and micropulsations do both exist in GRB, the differential energy spectrum of γ-rays must be steeper than $E_\gamma^{-2.5}$ in the energy range 0.1 MeV $\lesssim E_\gamma \lesssim$ 5 GeV.

3.4.2 Spectral features

The high resolution Ge detector on ISEE-3 recorded several interesting emission features in GRB 781119 (Teegarden and Cline 1980, 1981). The KONUS data base (Mazets *et al.* 1981*b, c, d*) revealed the existence of 30 absorption and 10 emission features among the 143 GRB reported. Three of them showed both the absorption and emission features (for example GRB 790622 shown in Figure 3.13*b*) which could merely be a chance coincidence (Hurley 1984). The number of cases showing features must in reality be higher; for, in weak bursts, poor statistics might be masking the effect. All the absorption features occur at low energies (30–70 keV), while nearly all the emission features are seen at energies greater than 400 keV. Both the absorption and emission features are known to evolve with time; this aspect will be taken up in Section 3.4.3.

3.4.2A Absorption features

Some examples of GRB showing absorption features in the KONUS data are given in Figure 3.13 (Mazets *et al.* 1980). Among the 30 GRB (from KONUS) showing absorption features, there was only one case, GRB 800419, which was qualitatively confirmed by the HXRBS experiment on SMM (Dennis *et al.* 1982). The reason for non-confirmation of other cases by other experiments might be that none of them had both a good energy resolution and a sufficiently low energy threshold (Hurley 1983*b*). Apart from this evidence from the KONUS experiments, there is one example of an absorption feature 55 ± 5 keV in the spectrum of GRB 780325, reported by Hueter and Gruber (1982) and seen with their HEAO-A4 detectors. Unfortunately there was no confirmation by others (this event occurred prior to the launching of the Venera experiments on GRB).

Most of the absorption features, centred in the region 30–70 keV, account for an absorption of 0.01 to 0.02 of the burst energy (Mazets *et al.* 1981*d*). The equivalent widths are in the range 4–30 keV. The authors interpret the absorption features as due to resonance cyclotron absorption of photons in the intense magnetic fields surrounding the source regions when their energy matches that of the first Landau level. The relation is given by

$$E_\gamma(\text{keV}) = heB/m_e c = 11.6B_{12} \tag{3.14}$$

where B_{12} is the magnetic field in units of 10^{12} G. Accordingly an absorption feature at 60 keV implies the existence of a magnetic field $\sim 5 \times 10^{12}$ G in the emission region.

Strictly speaking, the observed absorption features are not narrow lines. They are rather broad, having equivalent widths of 4 to 30 keV (Mazets *et al.* 1981*d*) corresponding to $\Delta E_\gamma / E_\gamma \sim 0.3$. The width may be a result of Doppler broadening or of the variation of magnetic field within the emission region. Doppler broadening of a line of energy E_γ is given by (Trumper 1978)

$$\Delta E_\gamma / E_\gamma = [8 \ln 2\,(kT/m_e c^2) \cos\theta]^{1/2} \tag{3.15}$$

where θ is the angle between the line of sight to the observer and the magnetic field. For typical values of $\Delta E_\gamma / E_\gamma \sim 0.3$ and $\cos\theta \sim 0.707$, $kT = 16$ keV, a value much lower than the temperatures (hundreds of keV) derived for thermal bremsstrahlung fits. One must postulate, then, that the cyclotron absorption takes place in a relatively cool layer that overlies the hot region which produces the continuum.

There does not appear to be any correlation between the line widths and continuum radiation temperatures (Teegarden 1982), which can be taken as further evidence that two different processes or regions are responsible for the line and continuum behaviour.

There is one example of cyclotron emission, namely GRB 790526, which shows up as a bump over the continuum at an energy 45 ± 5 keV

(Mazets *et al.* 1981*d*). In this case, the electron temperature must be comparable to or higher than the energy of the first Landau level.

Interpretations other than that of cyclotron absorption for the low energy features have been put forward by some authors. D. Q. Lamb (1982) has expressed the view that these features are probably the result of time averaging spectra in which the low energy cut-off varies rapidly. Bussard and Lamb (1982) suggest that the 10–50 keV absorption features are due to photoelectric absorption by heavy atoms in superstrong magnetic fields. Lasota and Belli (1983) show that dips could result from a superposition of two spectra: an optically thin thermal synchroton spectrum (with a turnover at 55–110 keV) and a softer black body spectrum. Mitrofanov *et al.* (1984) have wondered how the GRB 820320, which shows low energy absorption features – implying the existence of fields $\sim 10^{13}$ G in the source region – could also emit photons of up to 30 MeV (Rieger *et al.* 1982); such photons, which would pair produce against intense magnetic fields, are not expected to emerge from within the source regions.

3.4.2B Emission features

An example of an emission feature at 460 keV is shown in Figure 3.13*b*, taken from Mazets *et al.* (1980). According to Mazets *et al.* (1981*d*) the emission features usually occur in the energy range 330–850 keV. Most of them are at 400–460 keV and their energy content is 0.03–0.11 of that of the initial burst. These lines are interpreted as due to e^+e^- annihilations which are gravitationally red-shifted. The energy of the observed lines is given by

$$E_{\mathrm{obs}} = m_e c^2 (1 - 2GM/c^2R)^{1/2} \tag{3.16}$$

where M and R are the mass and radius of the source object. The observed red-shifts point to gravitational potentials $\sim(0.1 \text{ to } 0.2)c^2$ and are consistent with those expected at the surface of a neutron star. Due to the finite kinetic energies of the e^+e^-, the annihilation line energy may exceed 0.51 MeV (Aharonian *et al.* 1980). If such blue-shifts exist the observed line is a result of both the blue- and red-shifts. Observations indicate that the possible blue-shifts are not very large.

If the emission line widths ($\sim$250 keV) are entirely due to Doppler broadening, the corresponding electron temperatures, kT, are $\approx$15 keV (Teegarden 1982). Daugherty and Bussard (1980) and Mazets *et al.* (1981*c*) point out that the annihilation lines can also be broadened by the presence of a high magnetic field ($\sim 5 \times 10^{12}$ G).

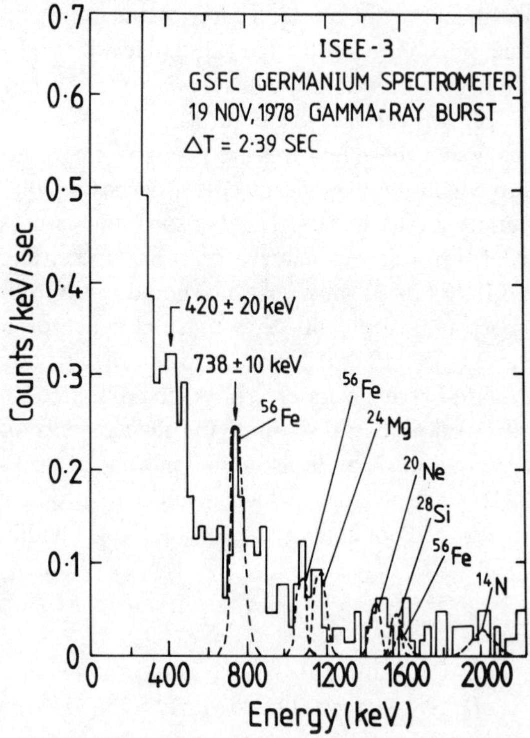

Figure 3.16. Energy spectrum of γ-rays from GRB 781119 obtained by the high resolution germanium spectrometer on ISEE-3 (Teegarden and Cline 1980, 1981). The feature at 420 ± 20 keV can interpreted as due to a gravitationally red-shifted annihilation line, the one at 738 ± 10 keV due to the red-shifted 847 keV first excited level of Fe and the others at higher energies due to red-shifted excitation lines from other nuclei.

The high resolution Ge detector on ISEE-3 produced the interesting spectrum from GRB 781119 shown in Figure 3.16 (Teegarden and Cline 1980, 1981). Although the feature at 420 ± 20 keV is termed 'marginal' by the authors, there is confirmatory evidence in the data base of KONUS experiments) see Figure 3.19 (Mazets and Golenetskii 1981*b*). SIGNE, too, confirm its existence over short time scales (Vedrenne and Chambon 1983). While the feature at 420 ± 20 keV can be interpreted as due to a gravitationally red-shifted annihilation line, that at 738 ± 10 keV is thought to be due to the first excited level of Fe (847 keV) red-shifted by the same factor as the 511 keV line. The other features in Figure 3.16 at higher energies could be excitation lines from other nuclei, red-shifted by the same factor.

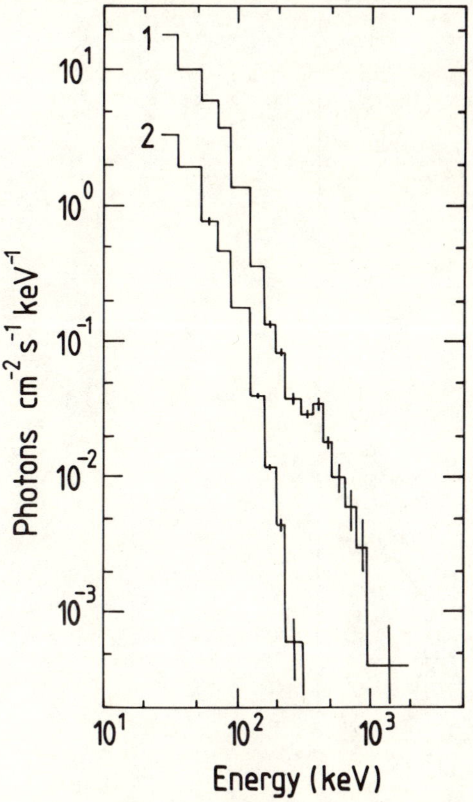

Figure 3.17. Evolution of the energy spectrum of γ-rays from the famous GRB 790305 as detected in the KONUS experiments (Mazets *et al*. 1981*a*). 1: spectrum recorded in the first 4 s interval following the onset, and 2: mean spectrum over the second to eighth 4 s intervals. Notice the disappearance of the feature at ~450 keV in the latter.

Mazets and Golenetskii (1981*a*) obtained evidence for the existence of an emission feature at 430 keV in the spectrum of the famous GRB 790305, which is shown in Figure 3.17. The authors interpret the feature as due to a gravitationally red-shifted annihilation line. Unfortunately the high resolution Ge detector on ISEE-3 was not usable at the time this burst occurred (Cline *et al*. 1981*b*).

Ramaty *et al*. (1980, 1981) explain the spectrum from GRB 790305 in terms of radiation originating in a thin ($\sim 10^{-2}$ cm) surface layer that overlies a hot plasma. In a magnetic field $>10^{11}$ G, cooling by synchrotron emission is much faster than by annihilation. Synchrotron emission from the rapidly cooling e^+e^- and the subsequent annihilation produce

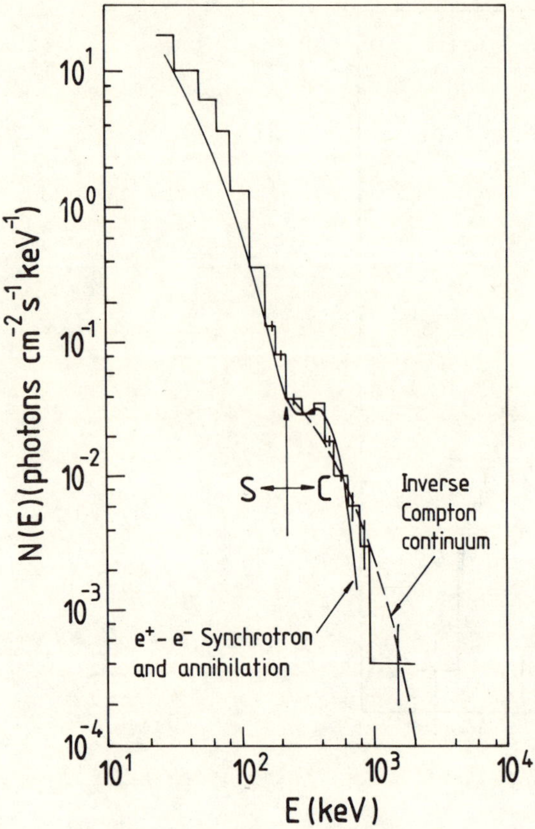

Figure 3.18. Theoretical fits to the energy spectrum of γ-rays from GRB 790305 obtained by Mazets *et al.* (1979*b*). The continuous line is the fit by Ramaty *et al.* (1980, 1981) on the basis of synchrotron emission by $e^{\pm}$ and the subsequent annihilation of the pairs. Dashed line improves the fit at higher energies (Liang 1981) by considering inverse Compton scattering of lower energy photons on ~MeV $e^{\pm}$.

the continuum and the line feature, respectively. This model has been modified further by Liang (1981) by including inverse Compton scattering from MeV e^+e^- pairs. The resulting spectral fits are shown in Figure 3.18. Liang (1981) also shows from first principles that the model can produce intrinsic luminosities as high as 10^{45} erg s^{-1} needed if the burst source was located 55 kpc away in the Large Magellanic Cloud (see Section 3.5).

There have been other interpretations proposed to explain the emission features. Ramaty *et al.* (1982), studying photon emission, absorption and scattering in a relativistic plasma of e^+e^- and γ-rays, show that an emission line at 430 keV can be produced by the GRASAR (γ-ray amplification through stimulated annihilation radiation) process, provided the pair chemical potential exceeds ~1 MeV. The authors stress that their GRASAR model operates in optically thick regions, thus obviating the necessity to have very large areas for emission regions, such as would be required if the emission were from optically thin regions. Katz (1981) has put forward an explanation for the 738 keV line seen in the spectrum from GRB 781119. At sufficiently high magnetic fields ($\geq 10^{13}$ G) the cross sections for one-photon annihilation far exceed those for the usual two-photon annihilations; see Figure 2 of Bussard and Lamb (1982). One would then expect to see 1.02 MeV ($=2m_ec^2$) lines. The ratio of energy of the one-photon emission line to that of two-photon emission lines as actually observed in this spectrum is 738/420 = 1.76 and not 2 as expected. Ramaty *et al.* (1982) point out, however, that the shifts in the annihilation peaks in their GRASAR model need not be by the same factor for the one- and two-photon peaks.

3.4.3 Variability of energy spectra

Variability of continuum spectrum and spectral features of a GRB is more the rule than the exception.The Franco–Soviet collaboration experiments, SIGNE, have demonstrated this variability of continuum and features on time scales of less than 250 ms (Vedrenne and Chambon 1983) and even on time scales as short as 30 ms in the case of GRB 790305 (Vedrenne 1984).

We have shown in Figure 3.19 the variability in energy spectra from GRB 781119 over time intervals of 4 s in the KONUS experiments (Mazets and Golenetskii 1981*b*). Notice how the bump around 400 keV, seen in the first spectrum, disappears in spectra taken a few seconds later and also the general softening of the continuum spectrum with time. One sees essentially similar time evolution in the spectra of GRB 791101 (Figure 3.13*c, d*), of GRB 790305 (Figure 3.17) and of several others. In particular, the famous GRB 790305 shows both soft ($kT \sim 35$ keV) and hard ($kT \sim 500$ keV) components in its impulsive phase (Figure 3.17) but only the soft component later on during the following oscillations (Mazets and Golenetskii 1981*b*) and also during its subsequent recurrences (Golenetskii *et al.* 1984). Spectral variations, as expressed by the associated changes in the temperatures of the fits for the three bursts GRB 790419, GRB 781104 and GRB 790113, are shown in Figure 3.20, taken from Teegarden (1982). In the first example, the temperature monotonically decreases with time; in the second it peaks well into the

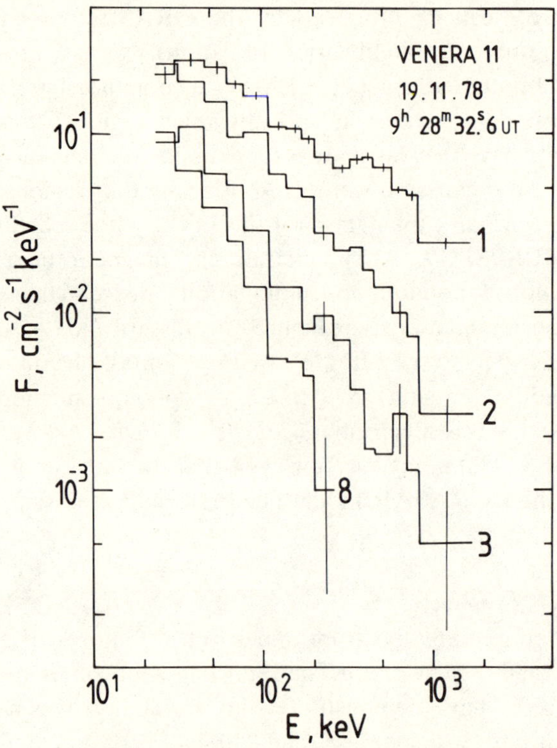

Figure 3.19. Energy spectra from GRB 781119 obtained in the KONUS experiments during the first, second, third and eighth 4 s time intervals from the onset of the burst (Mazets and Golenetskii 1981*a*). The feature seen by the Ge detector on ISEE-3 in Figure 3.16 at 420 keV can be seen as a broad hump here in the spectrum taken during the first 4 s; it disappears later. Notice also how the spectrum becomes softer with time.

burst; and in the third it exhibits ups and downs suggesting a positive correlation with the peak fluxes. Figure 3.21, also taken from Teegarden (1982), pertains to GRB 790110. It reveals a rich variety of phenomena; for example, at the beginning of the burst both the emission and absorption features are exhibited and these disappear later. The continuum which was fitted with an optically thin thermal bremsstrahlung form has a temperature $kT = 280$ keV to start with, increasing to 800 keV before it finally decreases to 440 keV. Time variability of the spectral features has also been seen by Barat *et al.* (1984*c*), who observed short duration line emissions from GRB 781104. The emission features were seen in the energy range 380–450 keV at confidence levels of 87–99 % in four

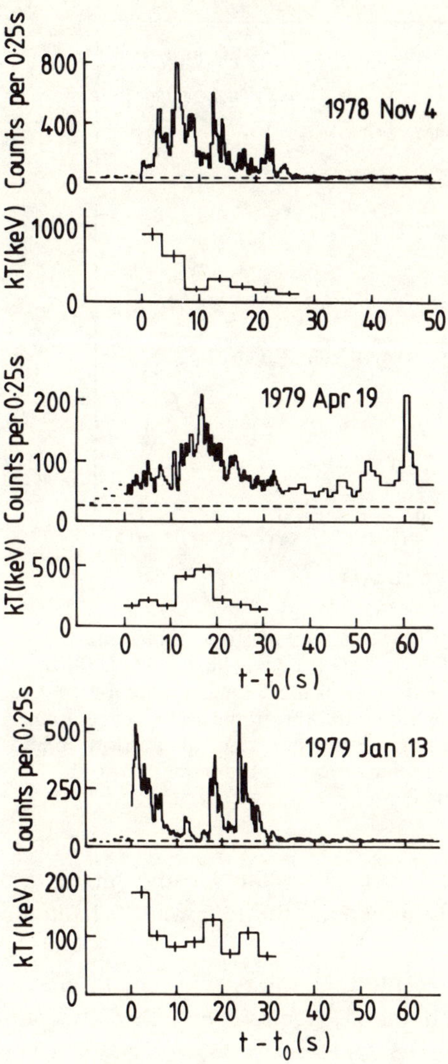

Figure 3.20. Time histories of total counts and spectral variations with time of γ-rays from three separate bursts GRB 781104, 790419 and 790113 as seen by the Leningrad group (Mazets *et al.* 1981*c*, 1982*a*). The figure suggests a positive correlation between the intensity and the characteristic spectral temperature, at least for some peaks.

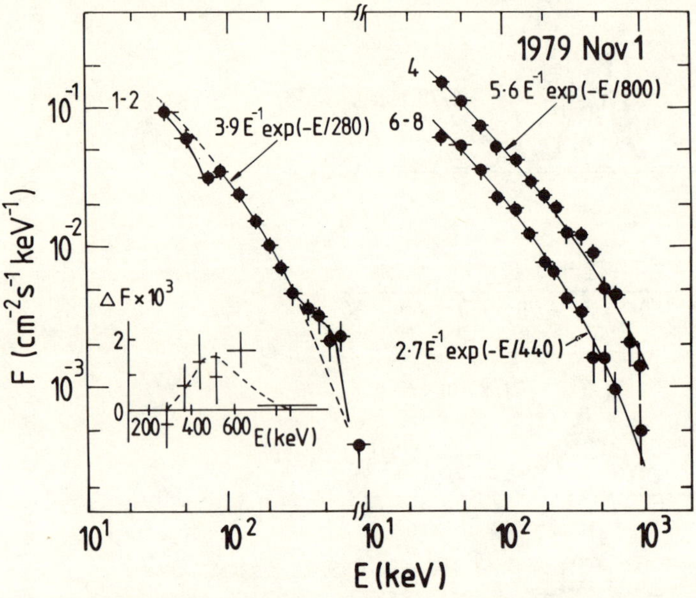

Figure 3.21. Evolution with time of the energy spectra from GRB 791101 from the KONUS experiments (Mazets *et al.* 1981*c*, 1982*a*) is shown. (1–2): the first 8 s of the burst; (4): the fourth 4 s interval; and (6–8): the combined data from the sixth, seventh and eighth 4 s intervals. Notice the disappearance of the absorption and emission features during the later stages of the burst as also the variation of the characteristic temperature with time.

different spectra, each integrated over 250 ms and mostly non-consecutive. The highest line emission flux among the four was 20.1 photons $cm^{-2}\ s^{-1}$.

Mazets *et al.* (1982*a*) have presented the interesting time profile and evolution with time of the continuum spectrum and the absorption feature of GRB 790731; see Figure 3.22. Here $\delta = (S_{tb} - S_{obs})/S_{tb}$, where S_{tb} and S_{obs} are the interpolated continuum spectral flux and the observed flux, respectively, at the absorption feature. Likewise the authors also show the time profile and evolution of the continuum spectrum and of the emission feature of GRB 790502; see Figure 3.23.

Spectral features whenever present are in most cases strongest during the initial phase and decay away much faster than the continuum. The continuum itself tends to exhibit a cooling down with time, although such behaviour is not universal (see, for example, Figures 3.20 and 3.21). These time variations present both an opportunity and a challenge to any

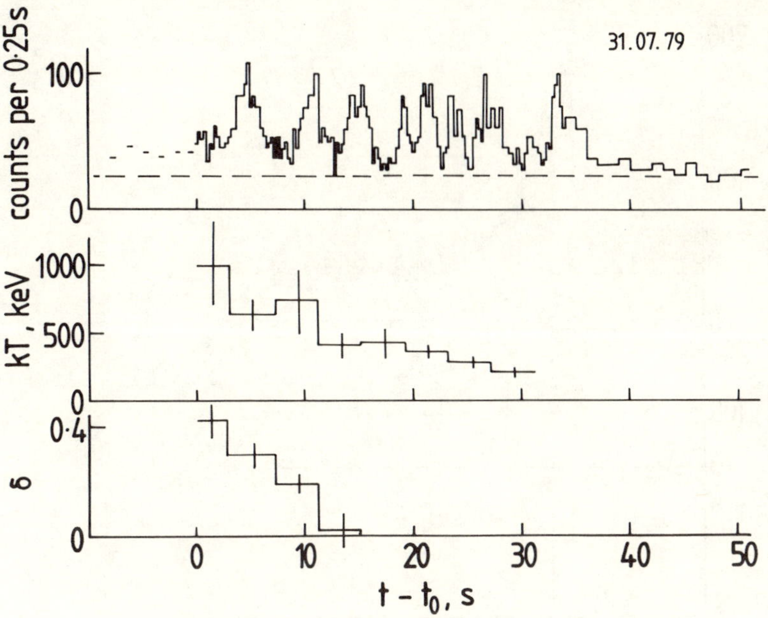

Figure 3.22. Time history of total count rate and variation with time of the temperature, kT, and of the amplitude, δ, of the absorption feature in GRB 790731 as seen in the KONUS experiments (Mazets *et al.* 1982*a*).

successful modelling of GRB production mechanisms and source objects and a proper understanding of their nature.

3.4.4 Correlations

It has already been mentioned in Section 3.3.1 that the burst fluences (integrals of fluxes over time) show a positive correlation with the corresponding burst duration. It is not clear, however, if most short duration bursts have soft spectra, as claimed by Mazets *et al.* (1982*b*); see the discussion in Section 3.3.1. Mazets *et al.* (1983) and Golenetskii *et al.* (1983) present evidence to show that the fluxes and temperatures of the continuum spectra of GRB are positively correlated. Due to an unintended drop of voltage across the photomultiplier in one of the six detectors on Venera 14, its energy sensitivity window became displaced to ΔE_γ = 150–700 keV, while on Venera 13 it remained at the original ΔE_γ = 40–180 keV. This enabled the authors to compare the count rates in the

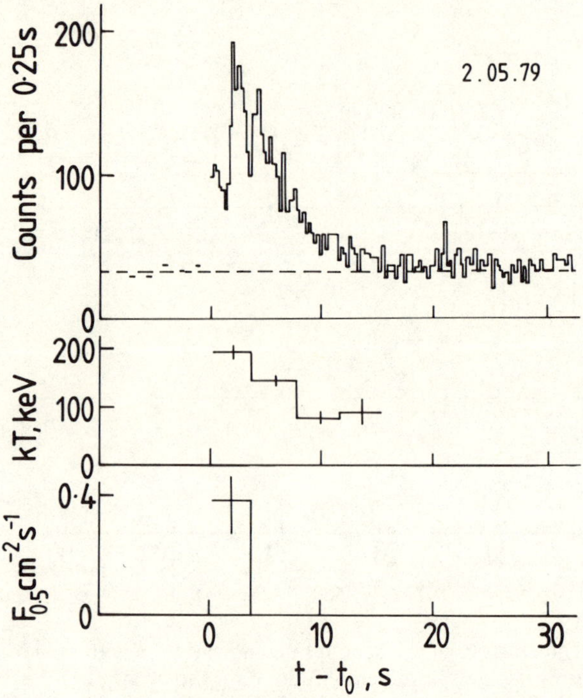

Figure 3.23. Time profile and variation with time of the temperature, kT, and of the annihilation flux, $F_{0.5}$, from GRB 790502 as seen in the KONUS experiments (Mazets *et al.* 1982*a*).

two energy windows with a 0.25 s resolution and obtain information on fast spectral variation. We have shown in Figure 3.24 count rates (N_{13}, N_{14}) for the GRB 820827 c as recorded by detectors on Venera 13 and 14, the ratio N_{14}/N_{13} and the temperature deduced from this ratio as a function of time. It is qualitatively apparent from the figure that the higher the count rate, the higher is the temperature. Results from a quantitative study of four bursts are shown in Figure 3.25. Fits of the type $L \propto (kT)^{\gamma}$, where L is the luminosity of the source (as reflected by count rate) and T is the temperature (as revealed by the spectral shape) have been tried. The data from GRB 820828 and GRB 820827 c are statistically the most significant and yield values for $\gamma = 1.70 \pm 0.15$ and 1.65 ± 0.10, with correlation coefficients of 0.94 and 0.93, respectively. γ values for all the six plots range from 1.5 to 1.7 for kT = 100 to 1000 keV. γ may vary from burst to burst and within the same burst in its individual peaks. This correlation suggests, then, that the time profile is directly determined by

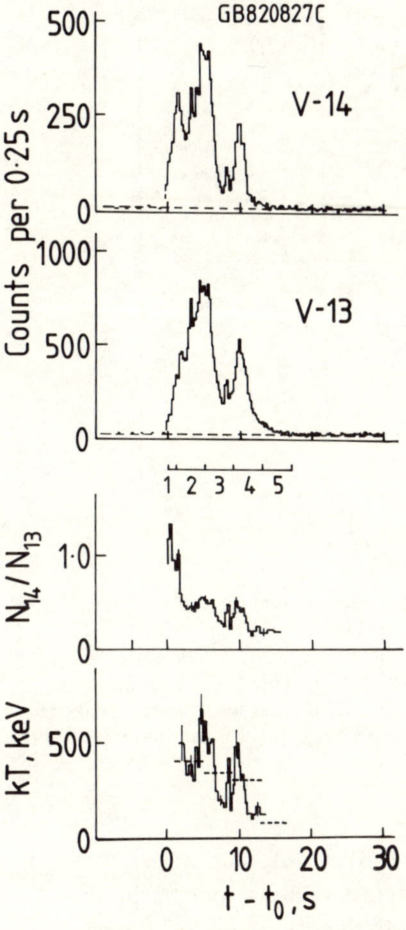

Figure 3.24. Time profiles and variation of hardness ratio, N_{14}/N_{13}, and temperature, kT, of GRB 820827 c as seen in the KONUS experiments (Mazets *et al.* 1983). Dashed lines in upper two panels: background counting rates. The temperature values, kT, are determined from the ratios, N_{14}/N_{13}, of count rates in detectors on Venera 14 and 13. Dashed lines in bottom-most panel: refer to temperatures of time-averaged (4 s) spectra during the intervals 2, 3, 4 and 5 shown in the third panel from the top.

spectral variability of radiation which, of course, has to have a deeper reason.

Since the luminosity is a function of temperature (kT), volume (V) and electron density (n_e) in the source region, the functional relationship $L \propto T^{\gamma}$ implies that either V and n_e remain constant or they vary with temperature in a rigidly controlled fashion for the duration of the burst.

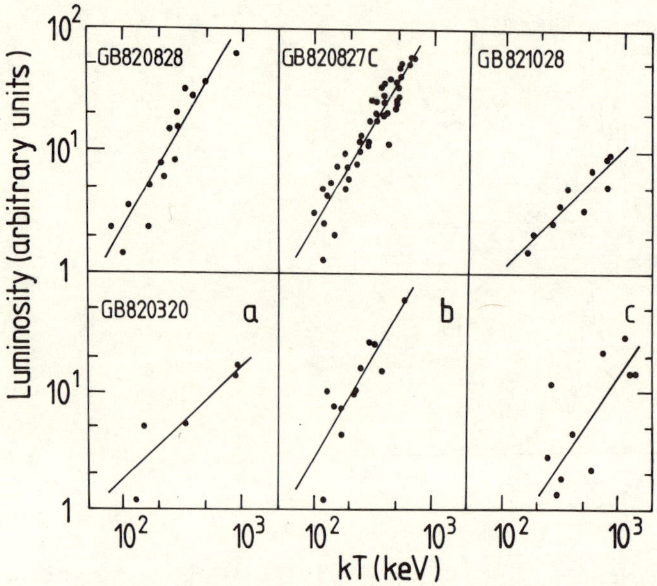

Figure 3.25. Correlation between luminosity and temperature in GRB inferred by the Leningrad group (Golenetskii *et al.* 1983). Instantaneous luminosity vs kT plots. Top: GRB 820828, 820827 c and 821028 events. Bottom: three stages (*a*), (*b*), (*c*) in the same multi-peaked burst GRB 820320. (*a*): for the weak initial peak; (*b*): for the decaying stage of the second peak; (*c*): for the initial part of the third peak.

Different production mechanisms predict different values for γ. For example, optically thin thermal bremsstrahlung models predict $L \propto n_e^2 V T^{1/2}$, which is in disagreement with the observed $\sim T^{1.7}$ dependence. Similarly, the authors conclude that the $T^{1.7}$ dependence is in conflict with the thermal synchrotron models of Liang *et al.* (1983) as well.

3.5 Identification of burst sources at other wavelengths

The optical identification of source objects has played a key role in the early development of radio and X-ray astronomies. Likewise the identification of GRB sources at optical and other wavelengths can be expected to be of crucial importance in establishing the nature of the sources responsible for GRB, in knowing if there is more than one type and in the successful modelling of GRB production processes. To do this, a precise localisation of a GRB source is a pre-requisite in order to avoid chance associations.

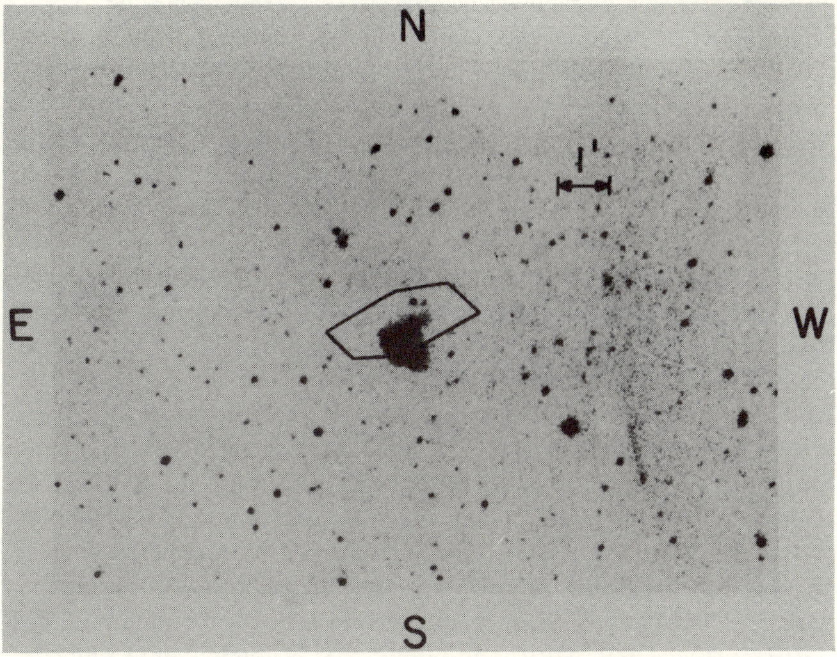

Figure 3.26. Association of the famous GRB 790305 with the remnant N49 in the Large Magellanic Cloud (Evans *et al*. 1980). The hexagon is the GRB localisation error box, and the irregular object near it is N49.

3.5.1 General description of methods

3.5.1A Catalogue searches

Knowing the GRB source direction, one simply looks up catalogues of interesting astronomical objects to find out if there are any objects in the error boxes. For example, Helmken *et al*., at the Centre for Astrophysics, Cambridge, USA, have compiled a catalogue of over 10^5 interesting objects such as pulsars, radio sources, nearby stars, white dwarfs, SNR, etc. (Hurley 1982). The association of the well known GRB 790305 with the N49 SNR in the Large Magellanic Cloud was quickly found by this method (Evans *et al*. 1980; see Figure 3.26).

3.5.1B Optical searches

Optical searches may be categorised as follows:

(i) *Standard sky survey plates* such as those from the Palomar Observatory Sky Survey plates may be searched for objects in the error boxes. In most cases the limiting magnitudes are $m \sim 21$. If found, these objects were probably photographed when they were in a quiescent state.

(ii) *Archival plate searches*. These searches, which assume that the GRB are repetitive, are made for transient emission of light that was presumably simultaneous with the occurrence of the burst in the past. The signature is the presence of a star-like object in just one plate but not in those taken earlier or later. The most famous example is the successful search for the source of GRB 781119 (Schaefer 1981); see Section 3.5.2.

(iii) *Real time searches*. In these one looks for optical emission in simultaneity with the GRB event. Grindlay *et al*. (1974) have argued that a reasonable extrapolation of the observed γ-ray spectrum to optical frequencies would predict optical magnitudes $m_v \sim 0$ for the source object. They have looked for time and spatial coincidences between GRB and optical flashes present in the records of Prarie Network of 16 stations meant to monitor meteor trails and, not finding any, have set a lower limit to the magnitude, $m_v \geqslant 6$, for the optical counterpart of GRB 720514. Deriving a ratio of the energy fluxes in γ-rays and in the optical window of >100, they conclude that GRB are not scaled-up versions of solar flares, where one expects this ratio to be ~ 1. Hudec *et al*. (1984), using the Czechoslovak Fireball Network, did not find any optical counterparts for GRB 781124, GRB 790107 and GRB 790325. London and Cominsky (1983) postulate a model in which γ-rays from a binary source are reprocessed by the companion star resulting in the emission of optical light. They predict that one can record as many as 100 events in a year, provided the instrument is sensitive to optical flashes at intensities ~ 200 photons cm^{-2}.

(iv) *Deep sky searches*. These are indicated whenever standard catalogue or survey plate searches do not show up any object (which is mostly the case) and when error boxes are known precisely enough either directly or through the intermediary of optical transients. If the search is deep enough, reaching a limiting magnitude $\gtrsim 25$, some object or other is bound to be seen in the error box, thus allowing one to state that no GRB error box is truly empty. At magnitudes $m_v > 22$, the density of galaxies equals or exceeds that of stars (Tyson and Jarvis 1979). This leads to a dilemma and the association cannot be relied upon because extragalactic objects, in most cases, are unrelated to GRB.

3.5.1C X-ray searches

These are of two varieties: (i) real time, and (ii) *post-facto*.

(i) Real time searches are carried out by looking at data from an X-ray detector operating contemporaneously (often on the same spacecraft) with a GRB detector. Examples of positive identifications of GRB at X-ray energies are the Apollo 16 event, GRB 720427 (Metzger *et al.* 1974, Gilman *et al.* 1980); GRB 720514 and GRB 740723 (Terrell *et al.* 1982). One must note that the field of view of an X-ray detector is much smaller than that of a GRB detector, leading to difficulties in establishing associations.

(ii) *Post-facto* searches for associations have been made by collecting new X-ray data from the known GRB directions, by using the imaging proportional counter (IPC) and the high resolution imager (HRI) on the Einstein Observatory (HEAO-B). In contrast to (i), one obtains in these searches information on the quiescent state of the GRB source.

3.5.2 Results from specific searches

GRB 790305. Localisation of the source was achieved on the basis of the International Network first by Evans *et al.* (1980) and more recently, and more accurately (~0.1 arc min^2), by Cline *et al.* (1982). The source direction error box contains N49, an SNR in the LMC (Evans *et al.* 1980); see Figure 3.26. The N49 region was earlier observed to emit at radio wavelengths (Mathewson and Clark 1973). Independent of the occurrence of GRB, the N49 region was also observed in soft X-rays by the Einstein Observatory eight days before and 38 days after the GRB occurrence, as a part of the X-ray survey of the LMC. With the moderate resolution of the IPC, the X-ray image of N49 could not be resolved and the authors conclude that the soft X-ray flux did not change between the two observations by more than 2.2×10^{-12} erg cm^{-2} s^{-1}. The high resolution imager could resolve N49 as shown in Figure 3.27; however, there was no point source seen, leading to an upper limit of 2×10^{-12} erg cm^{-2} s^{-1} in the energy range 0.5–4.5 keV. It is necessary to remark that this is a limit on the quiescent flux and not necessarily on the flux at the time of occurrence of the burst. The refined GRB error box lies within the X-ray diffused image of N49 but displaced by 15 to 45 arc s.

Recently the N49 region was subjected to a deep optical search by using a CCD camera at the 1.5 m Danish telescope. Several stellar components were seen inside the refined GRB error box (Hurley 1983*a*). Pedersen *et al.* (1984) have also monitored the source position of this burst for ~910 h using a high speed photoelectric photometer with a diaphragm 20×80 (arc s)^2, attached to a 50 cm telescope at ESO/La Silla. During the period October 5, 1983, to February 23, 1984, the authors had seen optical bursts on three occasions lasting for durations of 200 ms (on October 27, 1983),

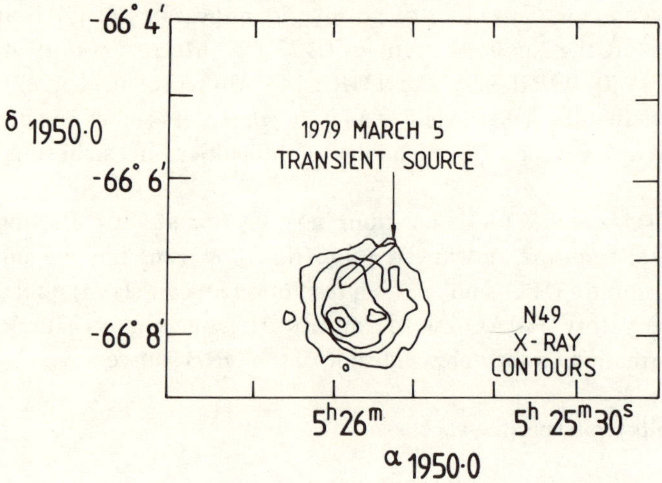

Figure 3.27. More precisely determined localisation error box for GRB 790305 (Cline *et al.* 1982) and the X-ray intensity contours of the supernova remnant N49 obtained by the high resolution imager on the Einstein Observatory. The GRB error box is displaced from the X-ray centre by ~15″–55″.

20 ms (January 30, 1984) and 400 ms (February 8, 1984). The γ-ray detectors on four spacecraft – ICE, PVO, SMM and Prognoz 9 – did not show any GRB simultaneous with these optical bursts. Neither did these detectors see any of the 15 recurrent bursts reported by the Leningrad group, which may be due to their threshold sensitivities being higher. It is not known if the more sensitive KONUS detectors on Venera 13/14 have seen any GRB coincidental with these optical bursts. If the association of the optical bursts with the source of GRB 790305 is real, the maximum brightnesses during the three optical bursts are $m_v = 8.4, 9.9$ and 8.7. The inferred values of the ratio of γ-ray to optical energy (L_γ/L_{opt}) during the bursts are less than 6000.

As mentioned in Section 3.1, association of this GRB with N49 has been questioned by Mazets and Golenetskii (1981*a*) and Golenetskii *et al.* (1984), mainly on the basis that the Eddington luminosity limit would be violated if the source were located as far away as 55 kpc; the source may therefore be a foreground object to N49. Felten (1981) considers the statistics of chance coincidence carefully and arrives at the conclusion that the chance coincidence probability for the association is less than 10^{-3}. Furthermore, Ramaty *et al.* (1980, 1981) and Liang (1981) have shown that there is no difficulty in devising models to provide the requisite

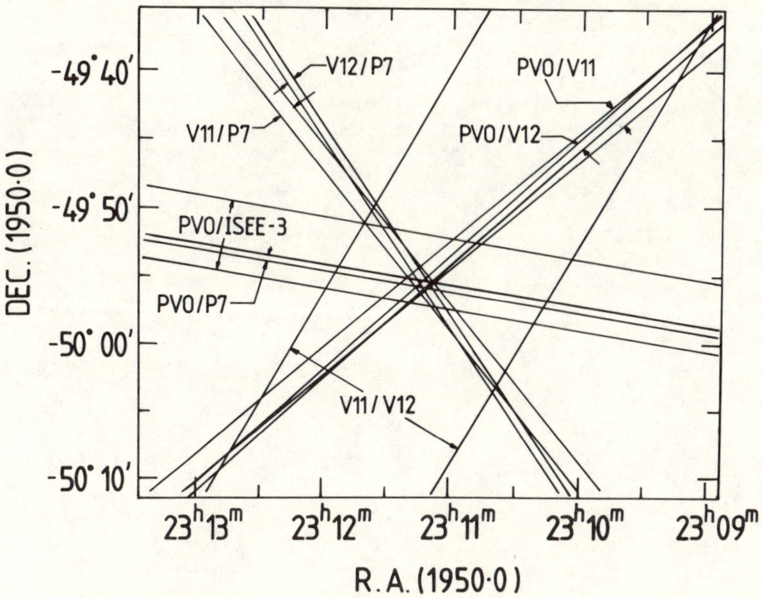

Figure 3.28. The error box determination for the GRB 790406 source direction as shown by Laros *et al.* (1981). The error bands in an approximately 0.5 square degree region centred on their common intersection are shown. The error box is defined by the PVO/V12, PVO/P7 and V12/P7 bands.

luminosity ($\sim 10^{45}$ erg s^{-1}) at the distance ~55 kpc; see Section 3.4.2B. As yet there is no consensus on the reality of association between the GRB 790305 and the SNR N49 in LMC.

GRB 790406. Figure 3.28 shows how this source was localised by combining information from several detectors (Laros *et al.* 1981). The error box is $<1'$ in diameter. Figure 3.29 shows the absence of any optical candidate down to $m_v = 22.5$ in the survey plates taken with the UK Schmidt telescope. The error box is also devoid of known radio, infra-red and X-ray sources. Five objects are seen in the error box in a deep sky search using a CCD camera with the 1.5 m Danish telescope of limiting magnitude 23.5; see Figure 3.30 (Hurley 1983*a*). It is not clear if these objects are really stars or galaxies.

GRB 790613. The source direction for this burst is localised to within 0.7 arc min^2 (Barat *et al.* 1984*b*). A search of 46 catalogues revealed

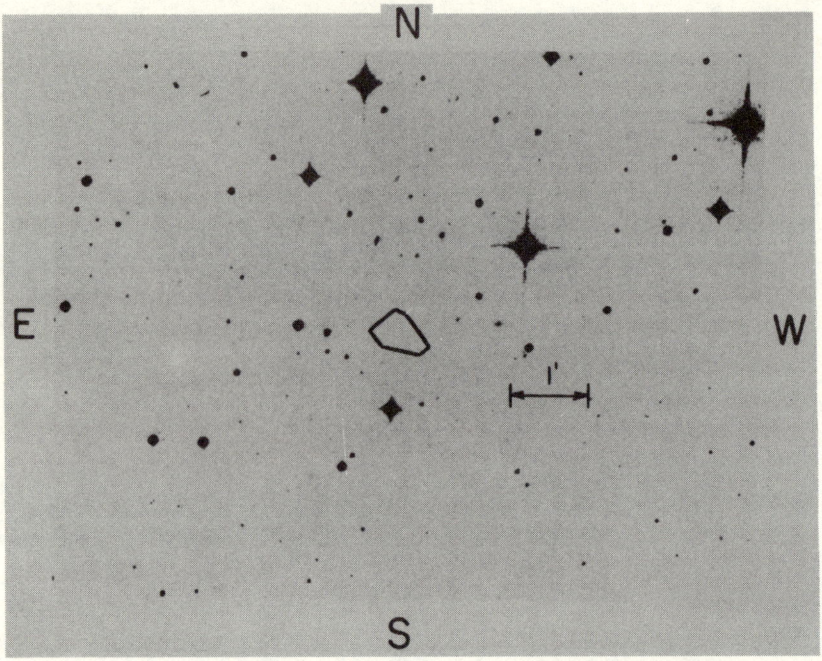

Figure 3.29. The figure shows the absence of any optical objects down to magnitude 22.5 in the error box (determination of which is illustrated in Figure 3.28) for the source direction of GRB 790406 (Laros *et al.* 1981). The photograph is from UK Schmidt plate 3659.

no object within the error box; nor did the Palomar Observatory sky survey plates show any object down to $m_v \sim 21$. Radio observations at VLA, too, did not reveal an object (Hjellming 1983). No optical transients were seen in archival plate searches (600 h exposure at Sonnenberg Observatory and 10 h at Harvard).

GRB 781119. Though the localisation error box for this burst was $\sim$10 arc min^2 (Cline *et al.* 1981*b*), the observation of an optical transient by Schaefer (1981), in an archival search in plates taken in 1928, within the GRB error box, effectively provided the link – one could substitute the much larger GRB error box with that of the optical flash which is only of size $8'' \times 18''$. The optical transient was found in the fourth of six identical 45 m exposures taken in succession on November 17, 1928. If a 1 s duration is assumed, the flash has a magnitude $m_B \sim 3$. Assuming that the 1928 and 1978 bursts were similar, Schaefer deduces that L_γ(>30 keV)/L_{opt} (B band) is 800. Since the source was seen in 1928 (optical) and

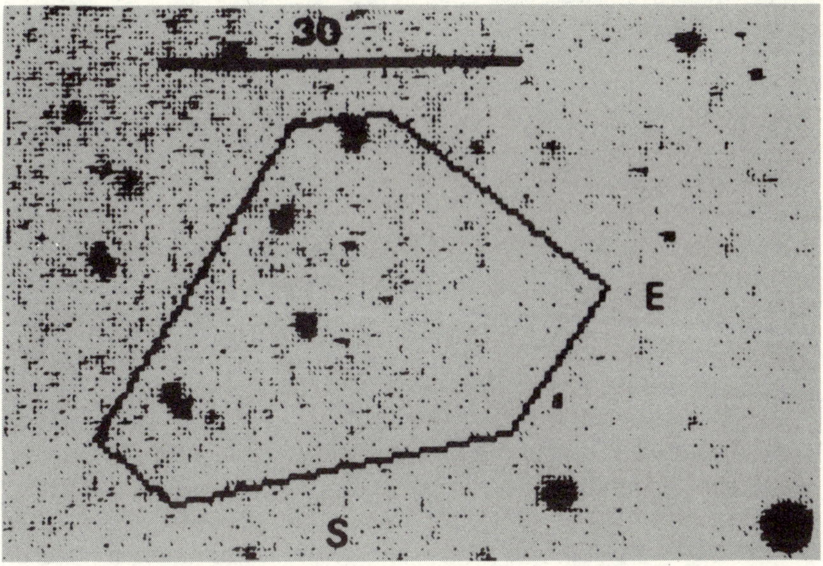

Figure 3.30. A partially processed CCD image of the error box region of GRB 790406 referred to Figures 3.28 and 3.29 (Hurley 1983*a*). The picture is taken at the Cassegrain focus of the 1.5 m Danish telescope at La Silla. Though five star-like objects (at $m \sim 23.5$) are seen inside the error box, it is not clear if these are stars or galaxies. If they are galaxies, they are probably unrelated to the GRB source.

again in 1978 (GRB), the recurrence time must be $\leqslant 50$ y. Hjellming and Ewald (1981) found several radio sources in and around the GRB error box with none of them coinciding with the transient error box. Grindlay *et al.* (1983), based on a 9000 s X-ray guest observation with the Einstein Observatory, found an X-ray source with an error circle of 45 arc s radius centred 42 arc s north-east of the optical counterpart; see Figure 3.31. The X-ray flux in the 0.5–3 keV band is 1.0×10^{-13} erg cm^{-2} s^{-1}, which is obviously the flux in the quiescent state. Since the quiescent optical flux seems to be variable (see below), the authors could set only a poorly determined limit to the ratio of fluxes $L_x/L_{opt} = 6$–60, which does not constrain the various models. The probable maximum luminosity is only $\sim 10^{31}$ erg s^{-1} and, according to Grindlay *et al.* (1983), supports the hypothesis that either GRB are produced by accretion instabilities or, if due to thermonuclear flashes, they recur at intervals of at least 50 y. Pedersen *et al.* (1983), on the other hand, conclude that, on the basis of the observed GRB fluence (3×10^{-4} erg cm^{-2}) and X-ray flux (10^{-13} erg

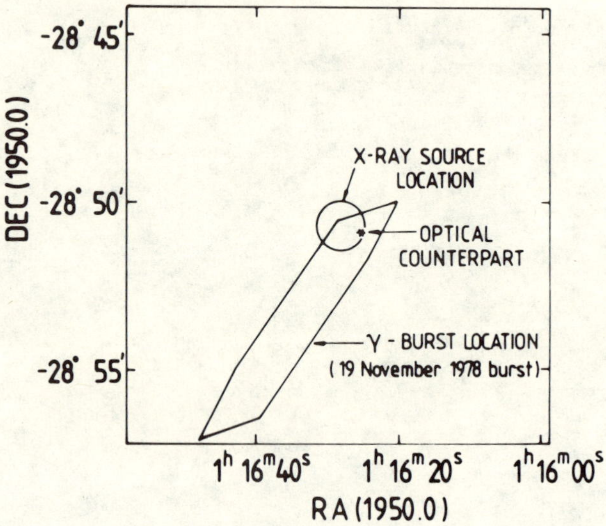

Figure 3.31. Location (90 % confidence error circle with 45 arc s radius centred at RA (1950) = 01 h 16 min 28.58 s, dec (1950) = −28° 50′ 42″) of the quiescent X-ray source which is the probable counterpart for GRB 781119. The GRB source localisation error box (Cline *et al.* 1981*b*), location of the 1928 optical transient position (Schaefer 1981) and the X-ray source error circle determined from the IPC detector on the Einstein Observatory by Grindlay *et al.* (1983) are shown.

$cm^{-2}\ s^{-1}$), either the conversion of accretion energy into X-rays is extremely inefficient or the thermonuclear models of GRB are not correct. In a deep sky search, Pedersen *et al.* (1983) found two objects A and B inside the optical transient error box; see Figure 3.32 and 3.33, at magnitudes $m_R \sim 23.5$. The authors found that object B was definitely variable by more than 1.3 magnitudes, while A was only mildly so. In an almost simultaneous observation, Schaefer *et al.* (1983) identified object A and two others inside the optical flash error box, but not B. These authors found that object A was variable by more than 1 magnitude over periods ~1 d. It is not clear if either A or B is the GRB source object. If neither is the source, then it is fainter than the 25th magnitude. Pedersen *et al.* consider the possibility that the source might have moved out of the transient error box in the 54 years from 1928 to 1982. Assuming a value of 1 km s^{-1} pc^{-1} for the velocity–distance ratio, the object could have moved out by 10 arc s. They found that there was no star-like image brighter than $m_R \sim 22$ within 25″ of the 1928 error box.

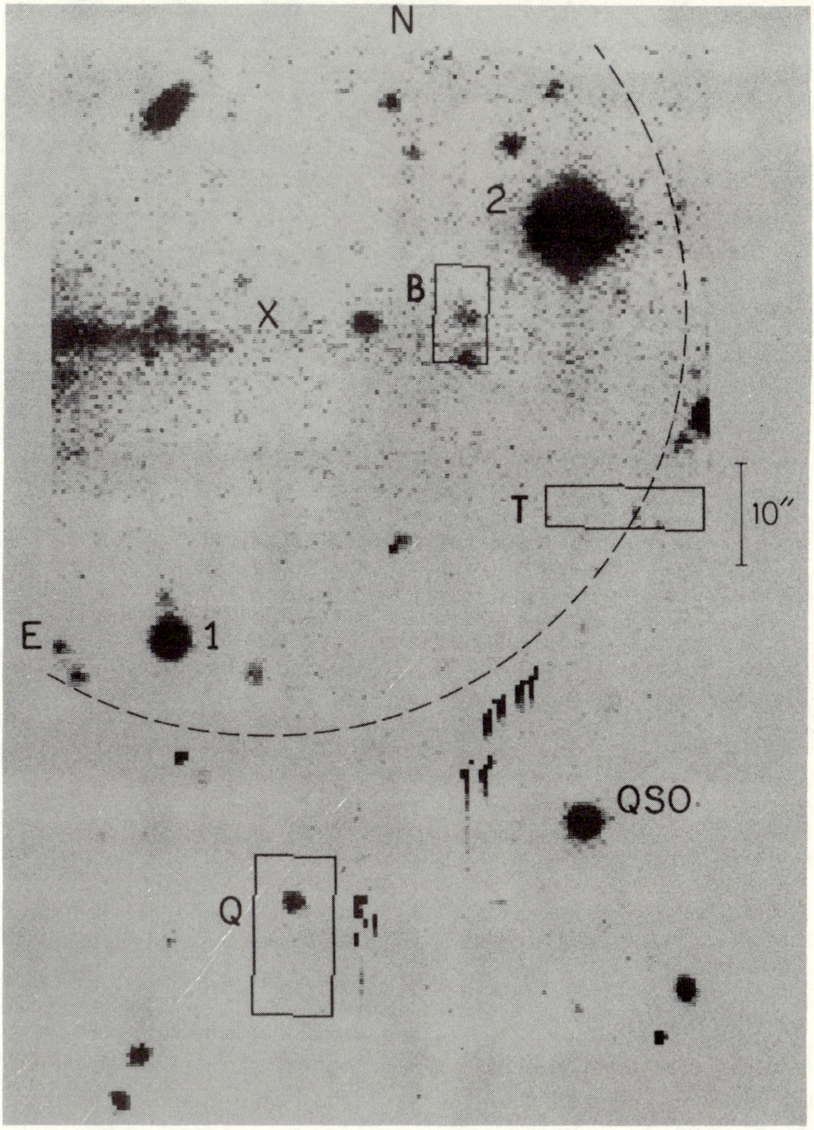

Figure 3.32. The region near the source direction of GRB 781119 is shown (Pedersen *et al*. 1983). The GRB source localisation error box, being much larger than the whole figure, is not shown. T is the 1928 optical transient error box (Schaefer 1981). X and circle: centre and error circle of the X-ray source (Grindlay *et al*. 1983; see Figure 3.31). B, Q: error regions of two radio sources (Hjellming and Ewald 1981). A QSO and two stars (1, 2) are also marked. Two suspected optical counterparts well within T are shown in Figure 3.33.

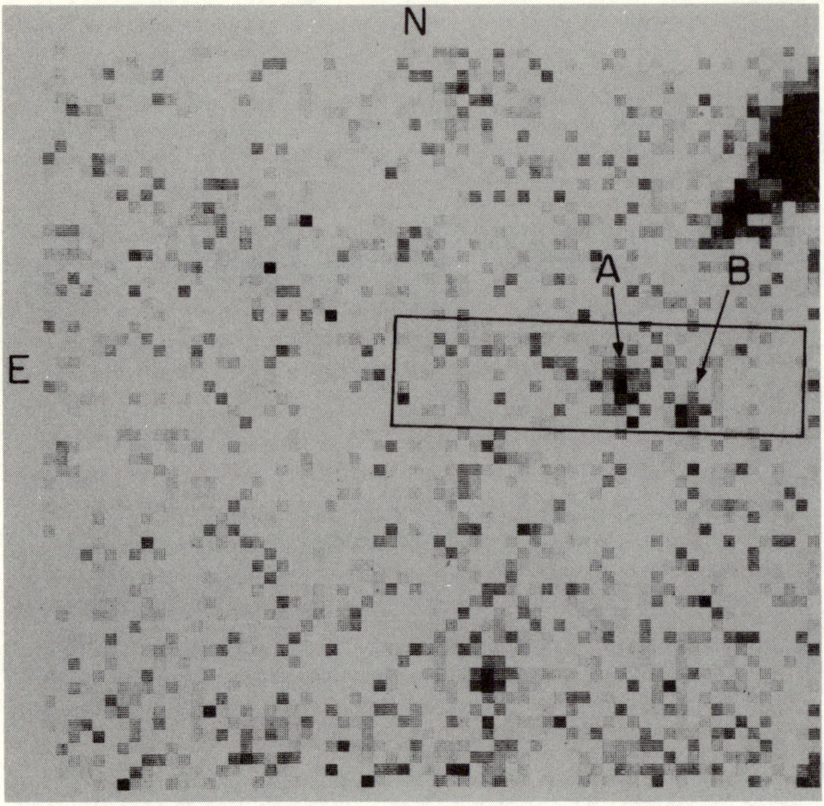

Figure 3.33. A close-up of CCD pictures summed over ~7.75 h exposure of the region near the 4″ × 16″ error box of the 1928 optical transient (Schaefer 1981, Pedersen *et al.* 1983). These workers find the optical object B ($m_R \sim 23.5$) to be variable by more than 1.3 magnitudes, while A is only mildly so. In a companion paper, Schaefer *et al.* (1983) reported finding object A to be variable and not seeing object B at all.

GRB 791105. Vedrenne (1984) states that Schaefer *et al.* (1984) detected in an archival plate search an optical transient in the photographic plates taken on October 4, 1901. If a 1 s duration is assumed for the optical flash, a magnitude of 6.6 could have been reached. The 1901 optical error box is empty in Palomar sky survey plates. The estimated value of L_γ/L_{opt} is of order 900, provided that the 1901 burst had the same γ-ray intensity as the one in 1979. The optical transient (1901) and GRB (1979) observations imply that the recurrence time is less than 78 y for this burst.

GRB 790113. The localisation error for this GRB is rather large

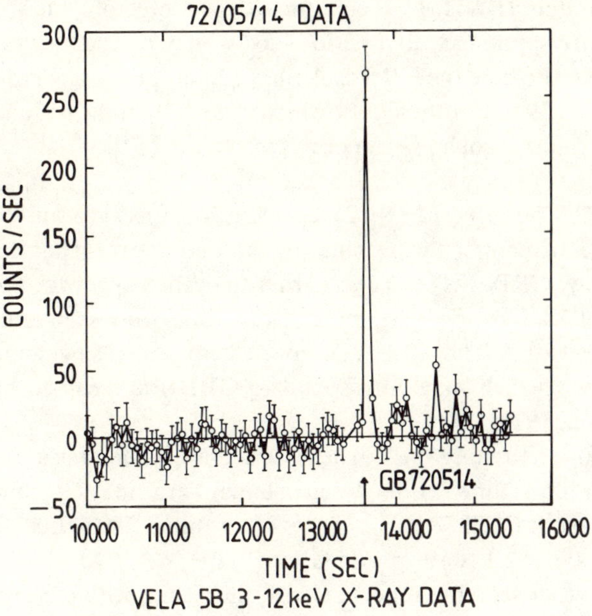

Figure 3.34. Time profile of X-ray counting rate in the 3–12 keV energy band as seen in the collimated X-ray detectors in the direction of GRB 720514 on VELA spacecraft by Terrell *et al.* (1982). The data are 3 s sums at 63.6 s intervals, corrected for the background. See the recurrence behaviour (smaller peaks to the right of the main peak) on time scales of hundreds of seconds. The GRB event time is shown by an arrow on the *x*-axis.

(~78 arc min^2) and several objects are found in the photographic plates taken with the ESO 3.6 m telescope (Barat *et al.* 1984*d*). Schaefer *et al.* (1984) found an optical transient in their archival search in plates taken in 1944. The error box of the optical flash is empty in ESO QB survey plates down to a magnitude ~21; however, more recent images taken with the ESO 1.5 m Danish telescope (limiting magnitude ~23) show clear evidence of an object in the transient error box. The estimated value for L_γ/L_{opt} is ≈800 (Vedrenne 1984). The recurrence time is less than 35 y for this burst, as deduced from the transient (1944) and GRB (1979) observations.

GRB 720514. An X-ray association with this GRB was reported by Terrell *et al.* (1982) using data from the collimated X-ray detectors (3–12 keV) on the VELA spacecraft. Near coincidence, both in spatial direction and in time (see Figure 3.34), establishes the identity. Since their view was limited (FWHM = 6°), the X-ray detectors on the spacecraft, which are spinning with a period ~64 s, may not be viewing the GRB precisely at the time the GRB took place. X-ray peaks from this object were first seen 39 s after the GRB, which was of 3 s duration, and four more times at intervals of a few hundred seconds.

GRB 740723. Terrell *et al.* (1982) report an X-ray association for this GRB in a similar manner to the burst mentioned above. The first X-ray peak from this GRB was seen 18 s earlier than the trigger for the GRB, which was of 7 s duration. Two more X-ray peaks were seen after intervals of 64 and 699 s. The VELA X-ray detectors also saw some isolated X-ray bursts which may correspond to GRB too weak to be detected (Evans *et al.* 1976).

Helfand and Long, searching the Einstein Observatory data base, set the following upper limits for X-ray emission in units of erg cm^{-2} s^{-1} in the 0.5–4 keV energy band: GRB 780305, $<2.1 \times 10^{-12}$; GRB 780329, $<2.1 \times 10^{-12}$; GRB 790406, $<2.1 \times 10^{-12}$; and GRB 791104, $<3 \times 10^{-13}$.

There are many other searches reported in the literature, with perhaps less significance than those listed above; see Hurley (1983*a–c*), Cline (1983) and Vedrenne (1984).

3.6 Statistics of gamma-ray bursts

Since GRB sources could not be identified with any known astronomical objects (except perhaps in the case of GRB 790305) an understanding of their origin is rendered difficult. Nevertheless it is possible to make a statistical study of some of the observed properties and hopefully get closer to the solution of their origin.

3.6.1 Frequency of occurrence

In the SIGNE (Franco–Soviet collaboration) experiments, detectors on Prognoz 7, Venera 11 and 12 yielded 40 GRB in 474 d, i.e. (0.08 ± 0.01) d^{-1}, during the period October 1978 to April 1980 (Hurley 1984). The same group later observed 104 GRB in 399 d, i.e. (0.26 ± 0.03) d^{-1}, in detectors on Venera 13 and 14. This three-fold increase in the detectable rate is attributed to a lower energy threshold and/or a more sensitive trigger algorithm in the latter detectors. In the KONUS experiments, 143 GRB were seen in 384 d, i.e. (0.37 ± 0.03) d^{-1}, in detectors aboard Venera 11 and 12. Later the KONUS group detected 180 GRB in 400 d, i.e. (0.45 ± 0.02) d^{-1}, in detectors on Venera 13 and 14.

Balloon experiments, though more sensitive in detecting GRB, are of short duration, and seldom more than one is detected in a single flight. Many groups could set only upper limits to the frequency of occurrence of GRB. The experiments of Bewick *et al.* (1975) and Beurle *et al.* (1981) suggest that the burst frequency is perhaps several thousand per year, i.e. $\sim 25\ d^{-1}$, at fluences $\geq 10^{-8}$ erg cm^{-2}. Helfand and Vrtilek (1983), using the data base of the Einstein Observatory, have set an upper limit of $10^5\ y^{-1}$, i.e $<274\ d^{-1}$, for GRB with fluences above $S \sim 10^{-10}$ erg cm^{-2} at $E_\gamma \sim 1$ keV.

If the emission of γ-rays is not isotropic, a certain fraction, depending on the degree of beaming, will go undetected even by the most sensitive detector. In such a situation, therefore, there is no sure method of estimating the true rate of occurrence of GRB in nature.

3.6.2 Source distances

A knowledge of the intrinsic luminosity of GRB is important in understanding the GRB production mechanisms. The observed quantities are the fluence and the source direction but not the distance, which is essential in calculating the luminosity. The circumstances prevailing at other wavelengths (e.g. red-shifts at optical wavelengths and the dispersion measure at radio frequencies) are simply not there to be utilised with the GRB. Consequently indirect arguments are brought to bear on the subject.

Noting that the spatial density of the GRB sources cannot exceed Oort's upper limit on the local invisible mass density, Schaefer and Ricker (1983) conclude that these sources (with assumed masses of $1.4\ M_\odot$) must be further than 3.2 pc from the solar system.

The peak intensity of the weakest confirmed GRB appears to be 2×10^{-7} erg $cm^{-2}\ s^{-1}$ from satellite data (Hurley 1984). If one assumes that the emission is isotropic, and the source is a neutron star of mass $1.4\ M_\odot$, the Eddington luminosity limit allows the source to be placed as far away as 2.5 kpc. However, there are situations where one need not be constrained by the Eddington luminosity limit.

GRB sources are isotropically distributed (see Section 3.6.3), and this implies that either the sources are very near (nearer than a few hundred parsecs), or very far (farther than the edge of the Galaxy). If the sources are far, they have to be beyond about 500 Mpc; for, at shorter distances, the distribution of matter is too inhomogeneous to produce isotropy. For example, the Virgo cluster ($\sim$18 Mpc away) would have introduced an

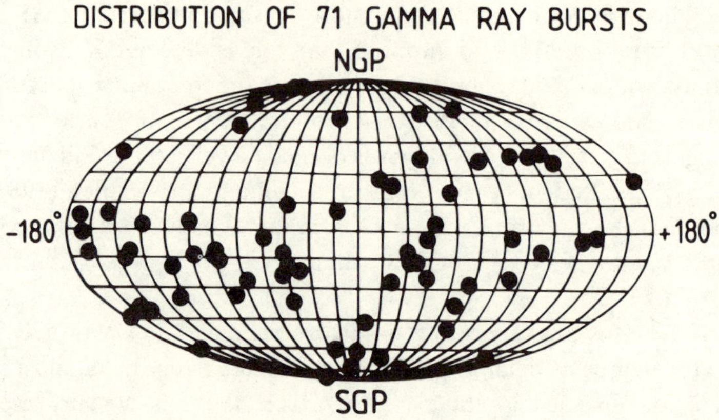

Figure 3.35. Distribution of source directions for 71 GRB in Galactic coordinates (Hurley 1983*b*).

unacceptable anisotropy (Jennings and White 1980). Such large distances would imply incredibly high luminosities and it is usual to fall back on the hypothesis that, by and large, GRB are Galactic objects.

Schmidt (1978) argues that, due to γ–γ interactions, there is an upper limit to the density of photons in the source regions emitting MeV γ-rays. Combining this with the information on time structure (to gauge the dimensions of source region) and observed intensities, he concludes that GRB sources cannot be further than a few kpc. Cavallo and Rees (1978) arrive at a similar conclusion. Liang (1982) points out that the absence of an observable low energy turnover due to synchrotron self-absorption puts an upper limit on the source luminosity and therefore on its distance at between 10 kpc and 1 Mpc.

3.6.3 Angular distributions

Source directions for 71 GRB are known with various accuracies (Hurley 1983*b*). These are shown in Galactic coordinates in Figure 3.35; the distribution is more or less isotropic. The distribution in latitude for the same GRB is shown in Figure 3.36. One sees an apparent excess of events over what is expected on the basis of isotropy at latitudes $\approx -15°$. Though the excess appears very striking, it is not statistically significant, being only a 2σ effect. Moreover, the latest KONUS data obtained with detectors aboard Venera 13 and 14 do not appear to show this anomaly (Mazets 1984). Also, there does not seem to be any correlation between the source intensity and the source latitude (Jennings 1982*a*).

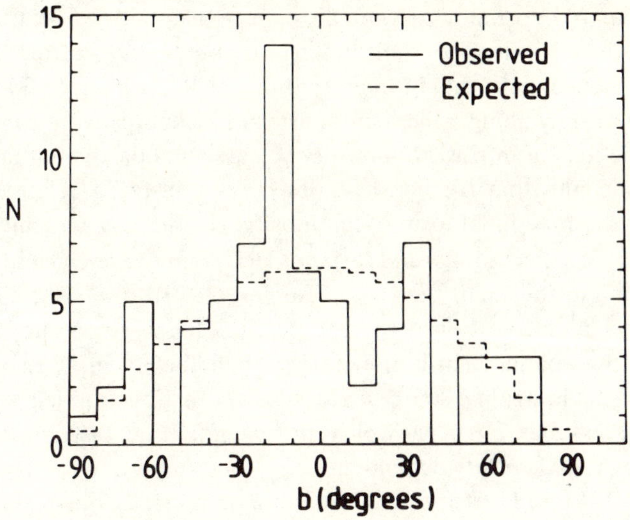

Figure 3.36. Distribution in Galactic latitude b of source directions for 71 GRB (Hurley 1983*b*). The expectation, shown by the dashed histogram, is on the basis of an isotropic distribution. Though striking, the anomaly at $b = -10°$ is statistically not very significant.

3.6.4 Log *N*–log *S* distribution

In these distributions, one plots $N(>S)$, the number of GRB with fluence, or size, greater than S, versus S. If the sources are uniformly distributed throughout space, one expects to see a relation of the type $N(>S) \propto S^{\delta}$, with $\delta = -3/2$, this form resulting from a cubical dependence of the number of bursts on radial distance (a constant density is assumed) and the inverse square law of burst intensity (monoluminosity is assumed) for isotropic emission. Likewise for sources confined to a thin sheet (approximating to the shape of the Galactic disc), one expects δ to be -1. For sources strung out along an infinite line (approximating the shape of the Galactic arm), on the other hand, $\delta = -0.5$. Since the Galactic volume, disc area or spiral arm length are all finite in extent, one expects the distribution to flatten over at sufficiently small values of S (below which there is no more contribution to the integral), and exhibit a $\delta = 0$ behaviour. One makes additional assumptions on the distribution of intrinsic luminosity of GRB sources, on the radial dependence in the case of halo models and on the height (above the Galactic Plane) dependence in the case of disc models.

Detailed mathematical computations have been carried out by Yoshimori (1978), Fishman (1979), Jennings and White (1980) and

Jennings (1982*a, b*). Yoshimori derives the expected log N–log S relation assuming that the burst density is uniform in space out to 0.3 kpc from the solar system, thence uniformly in the disc (Galactic Plane) out to 3 kpc and thence uniformly along a line (spiral arm) to 27 kpc. Having considered two functions of intrinsic luminosity (Gaussian and exponential), Yoshimori concludes that (i) the log N–log S relationship is relatively unaffected by the functional form of luminosity; (ii) there is agreement between data and expectations; and (iii) the GRB sources are Galactic in origin with luminosities $\sim 10^{39}$ erg s^{-1}. Fishman (1979), too, concludes that for a large number of observed sources, the log N–log S curve is governed by the spatial distribution rather than the intrinsic energy distribution of the individual sources. He rules out thin disc models with scale heights β less than 200 pc, as well as the extended halo distributions having a strong radial density dependence.

Jennings and White (1980), considering monoluminosity bursts, conclude that spherical halo models, particularly those with a central concentration, are unacceptable; so are thin disc models with a central concentration of sources. According to these authors, disc models with $\beta \geqslant 300$ pc are acceptable. Later, Jennings (1982*a, b*) generalises the model of Jennings and White to include an intrinsic luminosity distribution function of the form

$$n(L) = \frac{\alpha n_0}{L_2^\alpha(1 - \zeta^\alpha)} L^{\alpha-1}\,\mathrm{pc}^{-3}\,\mathrm{y}^{-1}\,\mathrm{erg}^{-1} \qquad (3.17)$$

Here n_0, the burst rate space density, is given by the relation

$$n_0(\mathrm{pc}^{-3}\,\mathrm{y}^{-1}) = \int_{\zeta L_2}^{L_2} n(L)\,\mathrm{d}L \qquad (3.18)$$

α controls the shape of intrinsic luminosity distribution and $\zeta = L_1/L_2$, the ratio of minimum to maximum luminosities. Disputing the conclusions of Yoshimori (1978) and Fishman (1979), that the intrinsic luminosity distribution does not affect the predicted log N–log S curve, Jennings demonstrates the vital role played by such a distribution, if it is of a proper mathematical form. Jennings concludes that both the disc and halo models are acceptable (see Figures 3.37 and 3.38, respectively) provided one chooses the right values for the parameters α and ζ. Parameters for the acceptable fits are shown in Table 3.1. According to Jennings, the soft luminosity distribution ($\alpha < +1$) fits explain why the predicted burst repetition (Jennings and White 1980) was not observed by early low sensitivity experiments. If the GRB are peaked at low luminosities,

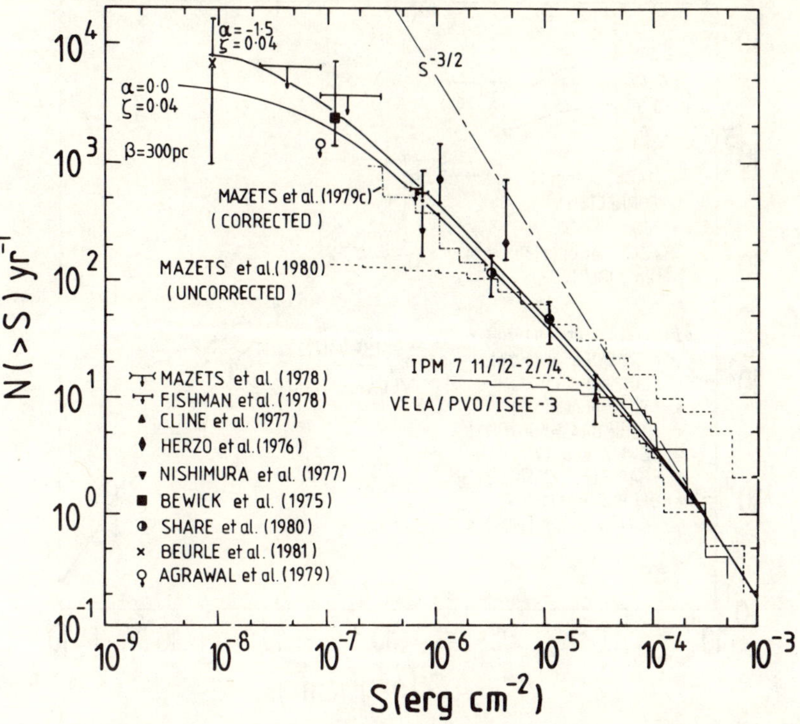

Figure 3.37. Log $N\,(>S)$ vs log S distribution of GRB. The histograms and points are from various experiments. Though shown, the data of Mazets *et al*. are not used in the fits. The two curves shown are fits by Jennings (1982*a*) with parameters $\alpha = 0.0$, $\zeta = 0.04$ and $\alpha = -1.5$, $\zeta = 0.04$ for disc models. In both cases, the half-thickness of the disc is assumed to be 300 pc.

Table 3.1. *Model fit parameters*

α	ζ	n_0 ($\mathrm{pc^{-3}\,y^{-1}}$)	L_2 (erg)	$\langle L \rangle$ (erg)	S^* ($\mathrm{erg\,cm^{-2}}$)
		Halo models			
−1.0	6×10^{-5}	5.2×10^{-10}	9.0×10^{42}	5.2×10^{39}	3.0×10^{-3}
−0.5	7×10^{-4}	1.9×10^{-10}	7.5×10^{41}	2.0×10^{40}	2.5×10^{-4}
		Disc models			
−1.5	0.04	9×10^{-9}	1.7×10^{40}	1.6×10^{39}	1.6×10^{-3}
−1.0	0.04	9×10^{-9}	1.2×10^{40}	1.6×10^{39}	1.1×10^{-3}
−0.5	0.04	9×10^{-9}	1.0×10^{40}	2.0×10^{39}	9.2×10^{-4}
0.0	0.04	9×10^{-9}	8.0×10^{39}	2.4×10^{39}	7.5×10^{-4}

* Luminosity below which log N–log S is affected.

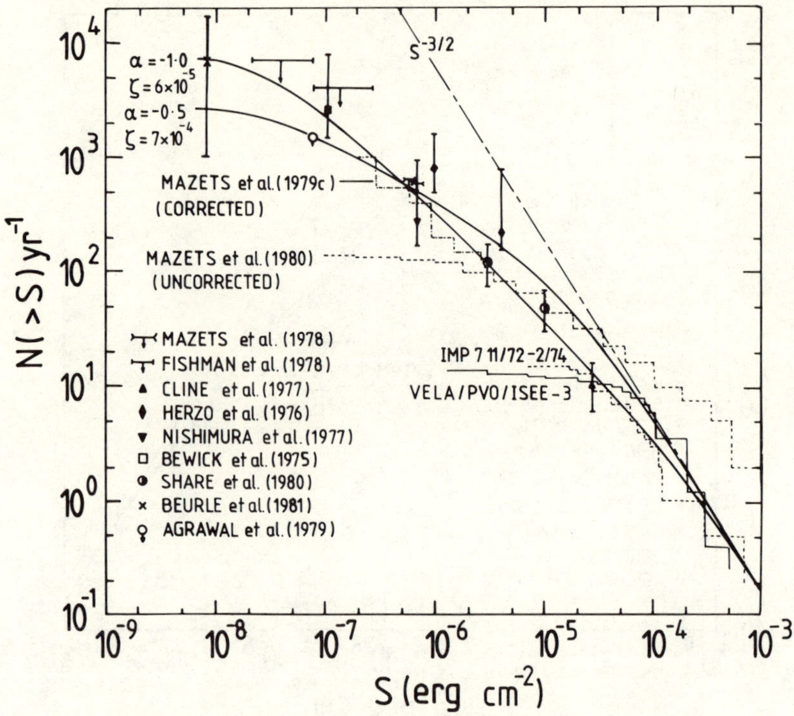

Figure 3.38. Log N ($>S$) vs log S distribution of GRB. The experimental data shown in Figure 3.37 are repeated here. The curves are fits by Jennings (1982*a*) with parameters $\alpha = -1.0$, $\zeta = 6 \times 10^{-5}$ and $\alpha = -0.5$, $\zeta = 7 \times 10^{-4}$. Both the fits are for halo models.

observations at low S assume increased importance, for these bursts are not simply geometrically diminished standard events more easily analysed at high S but are the bulk of the phenomenon itself.

Jennings (1982*b*) points out that the angular positions of the bursts and the fits discussed above are mutually inconsistent. For disc distributions, $\langle \sin b \rangle$ is a sensitive parameter for which Jennings calculates a value 0.33 based on the parameters shown in Figure 3.37, whereas the observed value (KONUS data base) is 0.465 ± 0.015. Similarly, a prediction for the ratio of bursts from the direction of the Galactic Centre to those from the Galactic Anti-Centre ranges from 1.6 to 2.5 for the halo models, whereas the observation shows 0.966. While one can contrive explanations for the inconsistencies between log N–log S and angular distributions, one wonders if the log N–log S curve contains *any* geometrical information.

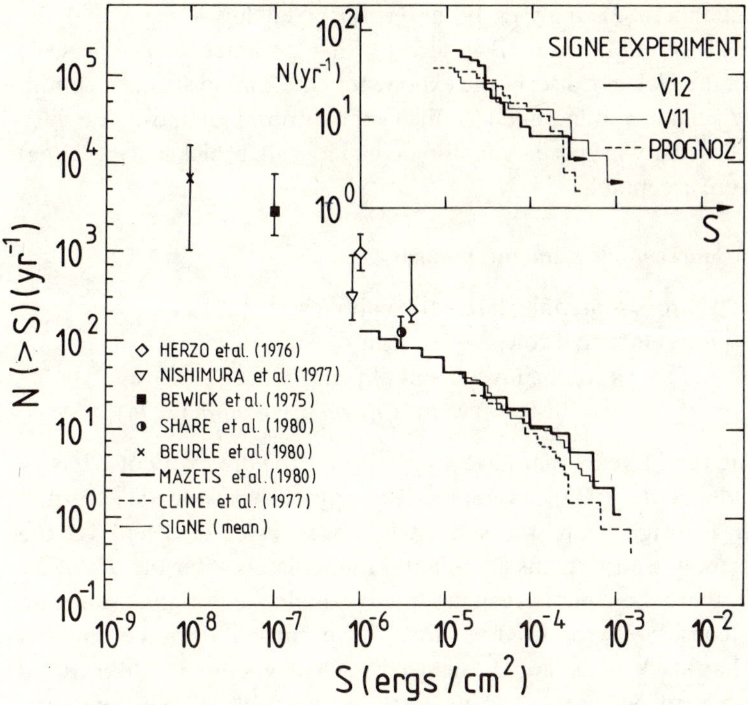

Figure 3.39. Log N ($>S$) vs log S curves from the experiments of Cline *et al.*, Mazets *et al.* and the Franco–Soviet collaboration (SIGNE) are mutually in agreement after correction for threshold and selection effects (Barat *et al.* 1982). The apparent discrepancies at $S > 2 \times 10^{-4}$ erg cm^2 are not statistically significant.

One notices that, though shown in Figures 3.37 and 3.38 for the sake of completeness, the log N–log S data of the Leningrad group were not used by Jennings in making the fits. The reason given is that they do not agree with other measurements and may have some internal inconsistencies. Barat *et al.* (1982) have shown that, if one calculates S over the same energy range with proper extrapolations, the observations made by different groups do mutually agree; see Figure 3.39. Klebesadel *et al.* (1984) have shown that if the log N–log S integral frequency distribution is plotted with steps at the level of fluences observed for each event, there is reasonable agreement between their data and those from the KONUS experiment. In the light of these two observations, the Jennings model (1982*a*) may need to be recalculated.

Chambon (1982) analyses both the observed log N–log S and the angular distributions of GRB in terms of five parameters: H, the scale height of the Galactic disc; α, the exponent in the luminosity distribution; L_1 and L_2 the lower and upper limits to the intrinsic luminosity; and n_0, the GRB density in space. Chambon concludes that thick disc ($\gtrsim$3 kpc) models are acceptable.

3.7 Source models and mechanisms

> 'There is something fascinating about science.
> One gets such wholesome returns of conjecture
> out of such trifling investment of fact.'
>
> Mark Twain, *Life on Mississippi* (1874)

During the 11 years that have elapsed since the discovery of GRB by Klebesadel *et al.* (1973), an impressive body of experimental facts and theoretical insights into the subject has been assembled and yet the phenomenon itself remains an enigma. Immediately after the discovery of GRB, there was a plethora of models to explain their origin. Candidate objects considered for GRB sources have ranged from Fermi-sized primordial black holes and 10 m-sized nuclear goblins to 10 km-sized accreting neutron stars. At the present time (1985), the consensus appears to be that GRB originate somehow on or near neutron stars with high magnetic fields. We will list these arguments in Section 3.7.1. The widely differing scenarios of energy release and the actual γ-ray production mechanisms, on which there is no consensus as yet, are briefly mentioned in Section 3.7.2.

3.7.1 Neutron star origin of gamma-ray bursts

The hypothesis that GRB originate on neutron stars with intense magnetic fields is supported by the following model-independent considerations:

(i) Fast variations in the time profiles of several GRB and periodicities observed in a few examples (see Section 3.3) strongly suggest that the source objects are very compact and dense, having densities greater than 10^9 g cm^{-3}.

(ii) Excess counts in the 400–500 keV energy range in the spectra of some of the bursts are interpreted as due to red-shifted annihilation lines (see Section 3.4.2B). This implies that the gravitational potentials in the emission regions are $\sim$(0.1 to 0.2) c^2. Such potentials are obtained near the surface region of a neutron star.

(iii) Deficit counts, in the 30–70 keV energy band, in the spectra of some of the bursts are explained as due to cyclotron absorption features (see Section 3.4.2A). This points to the existence of magnetic fields $\sim 5 \times 10^{12}$ G in the source regions. Such intense magnetic fields are believed to exist near neutron stars.

(iv) The maximum surface temperature (and hence the luminosity) that can be achieved by a neutron star in hydrostatic equilibrium occurs when the force of gravity just balances the radiation stress (Colgate and Petschek 1981; Ruderman 1981; D. Q. Lamb 1982), i.e.

$$\frac{aT_E^4}{4} = \frac{\mu m_H}{\sigma_T} \frac{Gm}{R^2} \tag{3.19}$$

Here a is the Stefan–Boltzmann constant; T_E is the maximum surface temperature; μ is the mean molecular weight per particle; m_H is the atomic mass of hydrogen; σ_T is the Thomson cross section; and M and R are the mass and radius of the star, respectively. While the resulting Eddington luminosities ($\sim 1.3 \times 10^{38}$ erg s^{-1}) can admit the observed GRB intensities up to a reasonable maximum source distance, the limitation imposed on temperature by the above equation, namely

$$T_E = 1.8\mu^{1/4}(M/M_\odot)^{1/4} R_6^{-1/2} \text{ keV} \tag{3.20}$$

is more restrictive. Here R_6 is the radius of the object in units of 10^6 cm. Observed energy spectra indicate temperatures ~ 200 keV, whereas equation (3.20) indicates very low temperatures, of the order of a few keV. If, on the other hand, a strong magnetic field is present, and if the plasma is radiation dominated, it can be confined perpendicular to the field provided the radiation pressure is smaller than the magnetic pressure, i.e.

$$\tfrac{1}{3}aT^4 \leqslant B^2/8\pi \tag{3.21}$$

This corresponds to

$$T \leqslant 170\, B_{12}^{1/2} \text{ keV} \tag{3.22}$$

Here B_{12} is the magnetic field in units of 10^{12} G. So magnetic fields $\sim 10^{12}$ G are indicated in the source regions to account for the observed spectra.

(v) Observed γ-ray fluences and various educated guesses on the source distances (see Section 3.6.2) imply that the total energy content of γ-rays is $\sim 10^{38\pm 2}$ erg. Neutron stars with intense magnetic fields are veritable energy reservoirs. For example, Brecher (1982) estimates that magnetic neutron stars have associated with them gravitational binding energies

$\sim 10^{53}$ erg, rotational energies (for periods $\sim$20 ms) $\approx 10^{49}$ erg and magnetic field energies $\sim 10^{44}$ erg.

(vi) Associations of GRB with optical objects are rare. If at all, the objects are very faint with magnitudes greater than 22 (see Section 3.5). Lone neutron stars are poor emitters of light. If they happen to be in binaries, therefore, their companions too have to be poor in emission of light. Ventura *et al.* (1983) outline the possibility of obtaining such low luminosity systems consisting of a neutron star primary and a degenerate star of mass $\sim 0.1 M_\odot$ as the evolutionary end product of a Galactic low mass binary system.

(vii) Periodicities observed in a few examples (see Section 3.3.3), besides supporting the argument given in (i) above, are natural for neutron stars. The well known GRB 790305 showed a periodicity with a period of 8 s, which is rather long considering the youth ($\sim$10 000 y) of the neutron star in the SNR N49. Brecher (1982) interprets this high period as that of precession and not of rotation, while Ruderman (1981) attributes it to the rotational period of a neutron star with an unusually high magnetic field ($\sim 2 \times 10^{14}$ G) at the pulsar. (Such is the latitude allowed by experimental facts for conjectures!)

It is not known if the neutron star is solitary or has a binary companion. Whether the energy source powering the burst is external to the neutron star or is internal to it is also unknown. It is significant that several other astrophysical phenomena invoke neutron stars for their explanation. These are: radio pulsars, X-ray pulsars, X-ray bursts and the lone example of the Jacobson transient. It is not clear if there is any relationship between any of them and GRB. In particular, X-ray bursts are seemingly very much unlike GRB in their energy spectra, repetition rate and Galactic distribution (see Vetter 1982). Magnetic fields and accretion rates appear to be key characteristics in some of the models. It is conceivable that variations in these characteristics lead to diverse phenomena; for example, Woosley (1982) thinks that low magnetic fields and high accretion rates onto neutron stars produce X-ray bursts, while high fields and low accretion rates result in GRB. Brecher (1982) is of the view that external energy sources (accretion) are responsible for X-ray burst production while internal energy sources (e.g. core quakes) produce GRB.

Any successful model of GRB origin must explain the fast risetimes, burst durations (few tens of milliseconds to few hundreds of seconds), fluences (10^{-8} erg cm^{-2} to several times 10^{-4} erg cm^{-2}), energy spectra, frequency of occurrence and periodicities and recurrences in the few cases that were seen. The models have also to confront the experimental constraints on energy ratios:

$$L_{XQ}/L_{opt\,Q} \sim 6\text{–}60 \tag{3.23}$$

(see Grindlay *et al.* (1983))

$$L_{\gamma T}/L_{opt\,T} \sim 800 \tag{3.24}$$

(see Schaefer (1981)) and

$$L_{XQ}/L_{\gamma T} < 10^{-9} \tag{3.25}$$

(see Helfand and Long 1979). Here the subscripts opt, X and γ refer to optical light, X-rays and γ-rays, respectively, and Q and T to quiescent and transient state of the source object, respectively.

3.7.2 Source model hypotheses

A large number of models for GRB production have been published. We will mention briefly only two of them, referring the reader to reviews by Prilutskii *et al.* (1975), Ruderman (1975), Vetter (1982) and Katz (1983) and to references therein for a more complete coverage.

As a prelude to a description of models, we may first mention the scale of energetics (Ruderman 1981). If the energy powering a GRB is derived from the gravitational potential ($\sim 0.1\,c^2$) at the neutron star surface, the available energy is $\sim 10^{20}$ erg g^{-1}. Likewise, the energy available from nuclear fusion ranges from $\sim 10^{18}$ erg g^{-1} (fusion of helium to heavier nuclei) to 10^{19} erg g^{-1} (fusion of hydrogen into helium). Typically, a GRB emits a total energy of $\sim 10^{38}$ erg in the form of γ-rays and the mass of material involved in the production of a GRB, then, should be $\sim 10^{19\pm 1}$ g ($\sim 10^{-14} M_\odot$).

Van Horn and Hansen (1974) first noted that hydrogen and helium accreted at the surface of a neutron star are subject to thermonuclear runaway for a wide range of accretion rates and surface gravity. This idea has been applied successfully to explain the production mechanism of X-ray bursts; see, for example, Joss (1978, 1979). A neutron star can accrete matter, mostly hydrogen and helium, from the stellar wind blowing out of its companion star or, more likely, from the overflow when the companion expands and fills its Roche lobe. Isolated neutron stars can accrete matter from the ambient interstellar medium. When the neutron star encounters dense clouds, the accretion rate may increase significantly. The accreted material may first form an accretion disc around the neutron star conserving its angular momentum and subsequently spiral

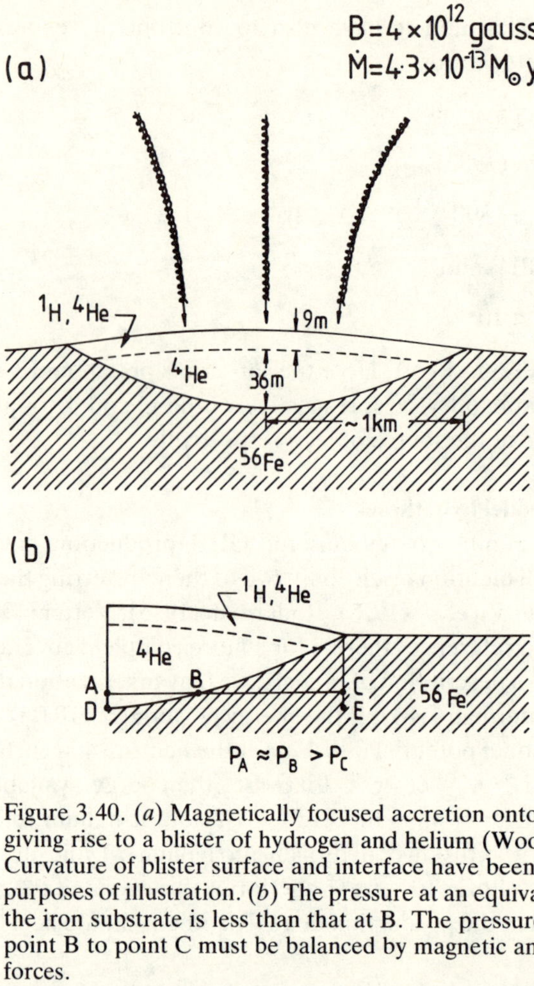

Figure 3.40. (*a*) Magnetically focused accretion onto a neutron star giving rise to a blister of hydrogen and helium (Woosley 1982). Curvature of blister surface and interface have been exaggerated for purposes of illustration. (*b*) The pressure at an equivalent depth, C, in the iron substrate is less than that at B. The pressure gradient from point B to point C must be balanced by magnetic and crustal shear forces.

down to the neutron star surface. Frictional forces due to differential rotation may heat up the material to a high temperature plasma, at which stage soft X-ray emission may take place. In the presence of magnetic fields, the plasma may be channelled to polar cap regions, increasing the local accretion rate by nearly three orders of magnitude compared with what it would have been otherwise for the case of isotropic accretion.

Woosley and Wallace (1982) consider an accretion rate $\sim 10^{-13}\, M_\odot\, \mathrm{y}^{-1}$ focused by magnetic fields $\sim 10^{12}$ G onto kilometre-sized polar regions of a neutron star; see Figure 3.40. The accreted material is confined to polar regions above the surface by the magnetic field and a combination of magnetic and crustal stresses below the surface. As the material accumulates, pressure builds up and, at appropriate pressures and temperatures, hydrogen burns in a steady fashion to form helium. When the helium reaches a critical mass, it explodes either by convective deflagration or detonation, thereby liberating 10^{38}–10^{40} ergs km^{-2} of thermonuclear energy. The plasma pressure at temperatures of the order of several billion degrees resulting from the explosion is greater than that of the magnetic pressure. The plasma, then, expands and stresses the magnetic field. Hard γ-ray emission comes from (i) relativistic electrons accelerated by large scale magnetic field recombination, (ii) magnetically confined and rapidly oscillating plasma, and (iii) particles moving with high velocities along the field lines in an optically thin region. The authors predict that soft X-ray emission should follow the GRB and endure for several minutes to an hour as the surface ashes of the thermonuclear explosion cool down. The model also predicts a direct correlation between burst energy and recurrence period.

Bonazzola *et al.* (1984) also subscribe to the view that GRB are powered by thermonuclear energy released at a huge optical depth, a few tens of metres below the surface of a neutron star. This extremely short phase is convectively unstable and, due to the high magnetic field, the convection has an oscillatory character. The thermal energy is then converted into magnetic energy which propagates very rapidly from the hot regions to the atmosphere of the accreted layer. Electric fields are created by short scale reconnection of the distorted magnetic field lines in the optically thin regions. Electrons are accelerated in these electric fields and produce the γ-rays.

In another scenario, magnetised neutron stars accrete solid objects (instead of tenuous matter) such as comets and asteroids of masses greater than 10^{17} g (Harwit and Salpeter 1973, Newman and Cox 1980, Colgate and Petschek 1981, Colgate 1982, Van Buren 1982). Occasionally a comet or an asteroid is deflected into an orbit with perihelion close to a neutron star. As the object falls in, the magnetic and gravitational fields distort and compress the body to densities $\sim 10^6$ g cm^{-3}. The interaction of matter with the field is initially entirely diamagnetic and the conducting matter parts the lines of force as a diamagnetic fluid. On impact, the object is in the form of a thin sheet of plasma, some 3 mm in longitude and 2.5 km in latitude; see Figure 3.41. After the collisional interaction with the neutron star surface, the matter and field become turbulent to inter-diffuse, decreasing the plasma pressure. Temperatures of $\sim 10^8$ K are released in a fireball and this is responsible for the GRB production.

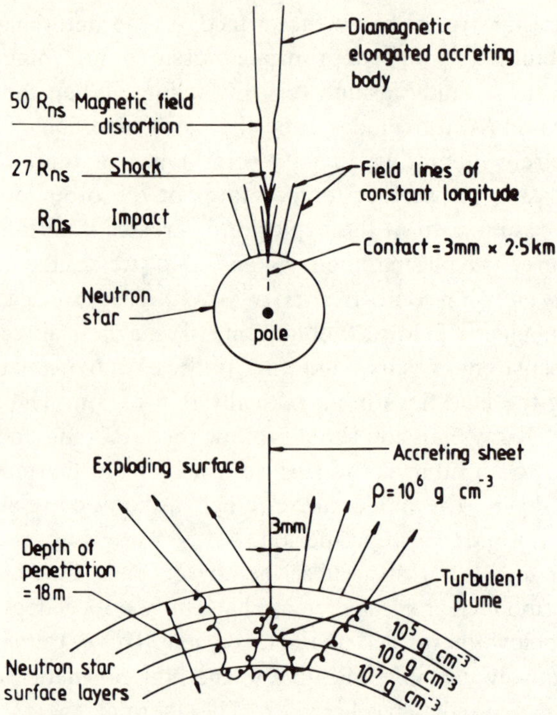

Figure 3.41. Accretion of a solid body onto a neutron star with a magnetic field (Colgate and Petschek 1981). The solid body, compressed and elongated, will be conducting and diamagnetic when it collides with a dipole magnetic field. It enters the dipole field as a diamagnetic compressed sheet of fluid. The body separates the magnetic flux surfaces of constant longitude and expands transversely in latitude. The compression in longitude reduces the thickness to a few mm and adiabatically compresses the matter to 10^6 g cm^{-3} density. It extends in width or latitude to a few km. The impact with the neutron star forms a turbulent mixed plume 20 m below the surface. The energy released by impact explosively ejects the combined surface and accreting matter.

Van Buren (1982) argues that in the fireball, with energy densities ~10 to 100 MeV nucleon^{-1}, e^+e^- pair production is ubiquitous. Consequently, spectral information which might have been a clue to the burst mechanism is lost since the fireball looks the same no matter how it was formed. Well before the impact, part of the parent body becomes plasma and falls freely to the neutron star surface while the asteroid orbit decays over a much longer time. The author suggests, therefore, that one should look for precursor radiation (duration ~10 s, luminosity ~10^{-5} of the main GRB event and having an optically thick type spectrum) which serves as a diagnostic of the actual mechanism of GRB production. For example, the author suggests, endogenic energy sources cannot produce such a precursor.

Many other models have also been suggested. Just to name a few, flare models (Katz 1982), crust quakes in neutron stars (Fabian *et al.* 1972, Bisnovatyi-Kogan and Chechetkin 1980) and core quakes in neutron stars (Tsygan 1975). An interesting GRASAR model by Ramaty *et al.* (1982) has been mentioned already in Section 3.4.2B.

Lamb (1982) pointed out that the burst durations are much longer than the cooling time of a hot plasma, whether in a strong magnetic field (synchrotron emission dominated) or not (bremsstrahlung dominated). In the accretion models, the burst duration is attributed (Colgate and Petschek 1981, Lamb *et al.* 1973) to the time interval during which accreting plasma or solid cometary matter rains down on the star. Nuclear burning models (Woosley and Wallace 1982) attribute it to the time necessary for a deflagration wave to propagate across the lake of nuclear fuel or the interval during which nearby fuels successively ignite. In either scenario, there is no difficulty in explaining the short risetimes of GRB.

Pederson *et al.* (1983) have remarked that the very low values of quiescent X-ray fluxes (a measure of accretion rate) associated with GRB are hard to understand unless, of course, the conversion efficiency, ε_X, of available energy into X-rays is very low. According to these authors, the recurrence time, τ_{recur}, of a GRB is given by

$$\tau_{recur} = RS_\gamma \varepsilon_X/(F_X \varepsilon_\gamma)\ \mathrm{s} \tag{3.26}$$

Here R ($\sim$100) is the ratio of the gravitational potential energy per nucleon to the thermonuclear energy per nucleon; S_γ (erg cm^{-2}), the fluence of GRB; F_X (erg cm^{-2} s^{-1}), the X-ray flux and $\varepsilon_\gamma(\sim 0.2)$ the efficiency of conversion of available energy into γ-ray energy. Applying these considerations to GRB 781119, one arrives at $\tau_{recur} \sim 1.5 \times 10^{12}\ \varepsilon_X$ s; compare this with the value $\tau_{recur} < 50$ y (1.5×10^9 s), derived in Section 3.5.2. Likewise, when applied to the consecutively recurrent bursts on March 5, and March 6, 1979, of the burst source GRB 790305, the formula predicts $\tau_{recur} \sim 1.5 \times 10^9\ \varepsilon_X$ s compared with $\sim$14.5 h time interval between the two bursts. These comparisons imply ε_X values of less than 10^{-3} and such values are hard to accept.

Woosley and Wallace (1982) predict that there should be a direct correlation between burst energy and recurrence period. This is not borne out by the several recurrences of GRB 790305 (Golenetskii *et al.* 1984), in which there is no correlation between the burst fluence and mean time between recurrences.

It is not clear if there is any understanding of the shape of energy spectra of photons from GRB. As detailed in Section 3.4, many of the observed spectra agree with an optically thin thermal bremsstrahlung model. In the presence of intense magnetic fields and at the temperatures believed to prevail in the source regions, the optical depth for synchrotron losses is greater than that for electron scattering by four orders of magnitude (Lamb 1982). Bremsstrahlung-like fits, though good, may not therefore reflect the actual mechanism of production of the γ-rays. Ruderman (1981) has shown that optically thick hot surface matter in a very strong magnetic field may also give a spectrum for the emitted radiation which departs greatly from a Planck distribution and which resembles the bremsstrahlung emission from an optically thin source. This is the sequel to the reduction of the photon–electron scattering cross section from σ_T in the presence of magnetic fields. See also the additional points made in the discussion presented in Section 3.4.1. The reality of an optically thin thermal bremsstrahlung spectrum is also doubted because of the uncomfortably large aspect ratios (i.e. ratio of $(\text{area})^{1/2}$ to thickness of the source region) that follow in such a case when one considers burst fluences, source distances and temperatures (Lamb 1982).

3.8 Summary

Gamma-ray bursts were discovered in 1973 by Klebesadel *et al.* (1973) and, during the past 12 years, the subject of GRB has received considerable attention, both theoretically and experimentally. To date several hundred GRB events have been fully analysed and perhaps a few hundred more have already been recorded; one awaits publication of full details of the latter events. With improvements in detector threshold and trigger algorithms, the rate of collection of GRB events has increased recently.

Most of the bursts have risetimes in the range 10–1000 ms, durations in the range of a few tens of milliseconds to few hundreds of seconds and fluences in the range 10^{-8}–10^{-3} erg cm^{-2}. The energy spectra can be fitted in most cases by an optically thin bremsstrahlung-like shape, although other forms of fits can be used. Indeed, in some cases, it is found necessary to use other types of spectral fit. The distribution of burst sources appears to be isotropic in Galactic coordinates. In two cases, the burst sources show recurrence and some of the bursts exhibit periodicity. A few of the bursts are strongly believed to be associated with objects that also emit optical light and X-rays, although these associations have not been confirmed yet in real time.

On the theoretical side, the consensus seems to be that GRB come from highly magnetised neutron stars. Models suggested for their origin include non-stationary accretion of material from the interstellar medium, from a binary companion or in the form of comets and asteroids. In some models, the power source is the thermonuclear explosion of accreted matter not far below the neutron star surface, near the polar regions. In other models the energy is derived from glitches and vibrations inside neutron stars. There is no consensus on what exactly is the power source. The precise mechanism of charged particle acceleration and subsequent γ-ray production is even less well established.

At the moment, data on GRB are being gathered by detectors aboard various spacecraft. In the future one wishes to see deployment of large area thin scintillators or xenon counters to increase the counting statistics. Large count rates will help establish the occurrence of precursors, post-burst activity and periodicity if these are of low intensities. The thin nature of counters will minimise errors made in de-convolution of energy-loss count rates in order to obtain true photon energy spectra. Usage of large area Ge detectors will enable one to record the emission and absorption features with high resolution and this will help, firstly to establish unequivocally the reality of the features and secondly to have an accurate measure of their parameters, such as intensity, width, etc.

It is not known if the burst size spectrum (log N ($>S$)–log S curve) saturates at low S values. Simultaneous balloon flights widely separated in distance but having a significant common solid angle of view will go a long way to establish the existence or absence of bursts at low values, $S \leq 10^{-8}$ erg cm^{-2}. It is also important to establish international networks of GRB detectors with well coordinated time-keeping methodology. This will help localise the burst source directions to $\lesssim$1 arc min which, in turn, will facilitate searches for associations at other wavelengths.

There is a clear need to establish if GRB are associated with any known class of celestial objects. Real time searches being planned at several places (e.g. the Explosive Transit Camera at Mt Hopkins, USA) and continual observation with an optical telescope of the sources of some of the well known bursts (e.g. GRB 790305) will be rewarding. With a mean recurrence time ~38 d (see Section 3.3.4) for the source of GRB 790305, the chances of seeing real time coincidental optical and GRB are really bright. Even a single observation of a real time association with an astronomical quality telescope would be extremely valuable. Detectors on many of the X-ray satellites may be able to see low intensity precursors or post-event activity associated with GRB.

New experimental missions that have been commissioned recently or likely to become operational soon (e.g. the International Solar Polar Mission, Hurley 1983*a–c*; the BATSE experiment on the Gamma Ray Observatory, Fishman *et al.* 1982; and the EXOSAT X-ray satellite, Anderson *et al.* 1984) will fulfil some of the needs listed above.

One can also expect studies of theoretical models to come up with predictions and suggestions on feasible experimental tests that are truly unique to the particular model, besides accommodating the burst features in general terms.

4

MEDIUM ENERGY GAMMA-RAYS

4.1 Introduction

For reasons concerned with the availability of contemporary γ-ray data, the lower limit for 'medium energy' quanta can be taken as 35 MeV (this is the lower limit for the important SAS II satellite experiment). The upper limit again comes from satellite data availability and is rather arbitrarily taken as 5000 MeV, the upper limit of the highest COS B satellite energy band; in fact, the photon flux falls off with energy so rapidly that our knowledge about γ-rays above 1000 MeV from satellite experiments is virtually nil. As will be discussed in Chapter 5, however, knowledge blooms again above 10^{11} eV, where Cerenkov radiation produced by γ-ray-induced electrons in the atmosphere allows detections to be made.

Although there are some who still believe that unresolved discrete sources contribute considerably, the majority view is that the sources are responsible for only 10–20 % of the γ-ray flux and that the predominant fraction arises from cosmic ray (CR) interactions with gas and radiation in the interstellar medium (ISM). In fact, some 30 years ago, both Hayakawa (1952) and Hutchinson (1952) had made estimates of the CR–ISM induced γ-ray flux and shown it to be within the scope of experimental measurement.

The foregoing is not to say that the discrete sources are unimportant, the reverse is true, and there is considerable interest in ways of explaining the observed γ-ray flux from identified sources (the CRAB and VELA pulsars) and from unidentified but definite sources such as Geminga (2CG 195+04 in the COS B source catalogue of Hermsen 1980, 1981).

The explanation of the diffuse component has considerable interest because of its rather obvious relationship to both the CR and the ISM, namely in regions where we think we know what the cosmic ray flux is, information can be gained about the ISM and in turn where the ISM is understood, the cosmic ray flux can be estimated. The need for information about the distribution of cosmic rays in the Galaxy cannot be overstated; despite the elapse of 70 years since the discovery of cosmic rays (Hess 1912) the origin of these particles is still the subject of debate and knowledge of the manner in which cosmic rays are distributed (local hot spots, large scale intensity gradients, etc.) is likely to give valuable information. The implication that the ISM itself needs further study is not as obvious but it is nevertheless true, particularly for the gas in the inner Galaxy. The situation here is that, although 21 cm data provide basic information about the gas densities of the atomic hydrogen, our knowledge of the important molecular component is remarkably poor. Molecular hydrogen densities at present are largely inferred from the measurements of the 2.6 mm emission line resulting from the $J = 1 \rightarrow 0$ transition in carbon monoxide (CO) and the factor for converting from the intensity of this line, or more specifically from $\int T \mathrm{d}V$ (where T is the antenna temperature and V is the velocity) to the column density of H_2, is the subject of lively debate today. The topic of interstellar chemistry, which can significantly affect the value of this conversion factor, is one of considerable complexity and uncertainty, largely due to a lack of precise knowledge of the physical properties of the medium where the chemistry occurs. Thus, γ-ray astronomy holds out the hope of independently 'weighing the Galaxy' in terms of mass of gas.

In what follows, after a brief discussion of the history of the subject, attention will be given to the two major satellite experiments, SAS II and COS B, which have given results in the medium energy region. This section is followed by a detailed examination of the Galactic γ-ray sources, leading in to the thorny but interesting question of the pseudo-sources, localised γ-ray excesses due to the irradiation of the lumpy molecular gas by cosmic rays. The relevance of γ-rays to the origin of cosmic rays follows next. The penultimate section deals with the rather sparse contemporary data on extragalactic γ-rays and, finally, new experimental programmes awaiting execution are discussed.

4.2 History

Pride of place for the first detection of γ-rays in this energy bracket appears to be due to Kraushaar *et al.* (1965) using the Explorer II satellite. The detector consisted of a low resolution (±20°) scintillation–

Cerenkov device and an important result that followed was the determination of the ratio of the earth's albedo : sky flux of the γ-rays of $\approx 10:1$. It was the secondary γ-rays produced by cosmic ray particles in the atmosphere that bedevilled (and still affect) the balloon-borne experiments and led to the first detection of genuine celestial γ-rays being reserved for the satellites.

It was in March 1967, with the launch of OSO III (Clark *et al.* 1968, Kraushaar *et al.* 1972), that the subject can really be considered to have started in earnest. The detector for the first time showed a broad peak in intensity towards the Galactic Centre of magnitude of the order of that expected from cosmic ray interactions with Galactic gas (summaries of the OSO III results have been given by Fazio 1970 and Clark 1971). These observations, made with a detector having a resolution of about $\pm 15°$, gave an average integral intensity above 100 MeV of $(1.1 \pm 0.3) \times 10^{-4}$ $cm^{-2}\,s^{-1}\,rad^{-1}$ for $30° > l > 330°$ and $|b| < 15°$. Clearly it was necessary to improve the angular resolution to see, for example, the extent to which the γ-ray intensity mirrored the column density of the gas (only atomic hydrogen, HI, was usually considered at that time) and a number of balloon experiments quickly followed. The peak towards the Galactic Centre necessitated balloon flights in the southern hemisphere and those of the Goddard Space Flight Centre group (GSFC) over Australia were successful. The detector adopted comprised a wire spark chamber with a magnetostrictive readout and an angular resolution of about 2° at 100 MeV (Kniffen and Fichtel 1970, Fichtel *et al.* 1972). The instrument is historically important in that it was the prototype of the SAS II satellite experiment, to be described shortly. The results showed that the bulk of the Galactic Centre direction ($335° < l < 20°$) radiation was confined to $|b| \leqslant 7°$ and the 'line intensity' above 100 MeV: $(2.0 \pm 0.6) \times 10^{-4}\,cm^{-2}$ $s^{-1}\,rad^{-1}$ was consistent with the earlier OSO III observation. The GSFC experiment also searched for point sources but was unable to find any at a level of $3 \times 10^{-5}\,cm^{-2}\,s^{-1}$ above 50 MeV.

In this period a number of other balloon experiments were also carried out, some of which saw a Galactic Centre peak (e.g. Share *et al.* 1973), whereas others did not, the problem being the short exposures available from the balloon flights and the considerable atmospheric background (see Share 1973 for a useful summary).

Although balloon experiments have continued to take place from time to time, and some useful data have accrued on specific 'sources', the bulk of the results have come from the two satellite experiments, SAS II and COS B.

The early results spawned many theoretical analyses and, of course, these are continuing. Before closing this historical section attention must be drawn to just one of the 'theorists', F. W. Stecker, whose early appreciation of the importance of γ-ray astronomy – and its relationship to other branches of astronomy – provided considerable impetus (see, for example, his published PhD thesis: Stecker 1971).

We turn now to a discussion of the satellite experiments.

4.3 The SAS II satellite

4.3.1 The experimental details

Insofar as considerable attention will be paid later to an analysis of data from this 'small astronomical satellite' (SAS) the experimental arrangement will be described in a little detail. Figure 4.1 shows the spark-chamber telescope carried by the satellite; the various components being labelled. Fuller experimental details have been given by Derdeyn *et al.* (1972), Kniffen *et al.* (1974), and Fichtel *et al.* (1975). The experiment was launched on November 15, 1972, and the orbit was essentially equatorial and roughly circular at a height ranging from 440 to 610 km above sea level. An electronic failure resulted in a premature termination of the experiment after only seven months operation, but nevertheless high quality low background data were accumulated. The 8000 photons

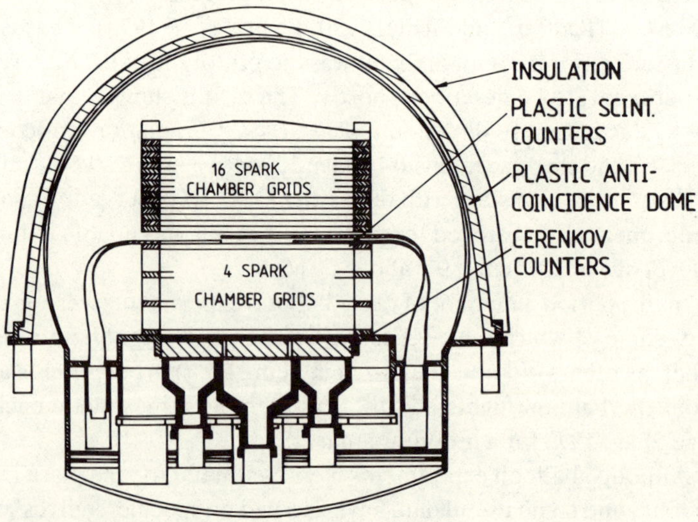

Figure 4.1. Schematic diagram of the SAS II γ-ray experiment (Derdeyn *et al.* 1972).

observed are still providing useful material for investigation some 12 years after the project was completed. In comparison with the early OSO III experiment, the SAS II sensitivity was about 12 times as great and its angular resolution reduced to a few degrees.

The data were recorded at a 1 kbit s^{-1} rate on redundant, continuous-loop tape recorders on board the satellite and the records were transmitted (at 20 kbit s^{-1}) once per orbit to a tracking station near Quito. The event epochs were recorded with an accuracy of better than 2 ms in absolute time.

The sky coverage of the experiment, which is an important parameter for analysing the results, is shown in Figure 4.2. It will be noted that, although the pointing direction was usually in the general region of the Galactic Plane, a significant exposure occurred for high latitudes as well, particularly in the northern hemisphere.

Energy calibration over the range 30–150 MeV was made prior to the flight using the tagged-photon facility at the 170 MeV electron synchrotron at the National Bureau of Standards (Gaithersburg, Maryland). Later analysis, involving folding in the energy uncertainty with an assumed incident energy spectrum of the form $N(E_\gamma)\,\mathrm{d}E_\gamma \propto E_\gamma^{-2}\,\mathrm{d}E_\gamma$,

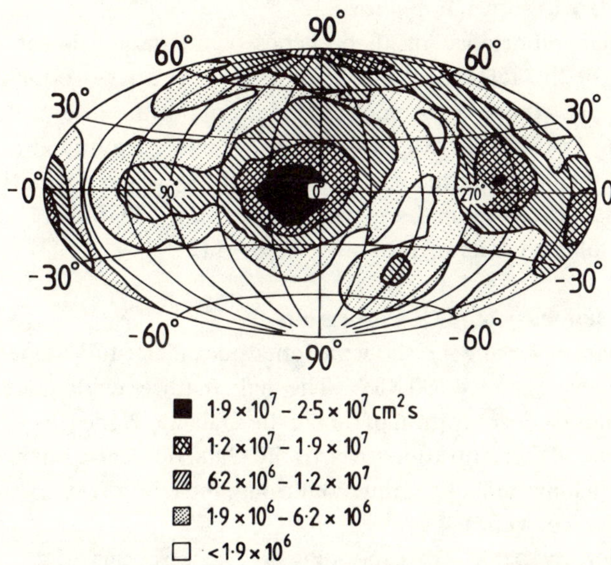

Figure 4.2. Exposure of the SAS II experiment (Fichtel *et al.* 1978). The horizontal axis is Galactic longitude and the vertical is Galactic latitude. The patchiness is due to the unexpected demise of the apparatus after only seven months in orbit.

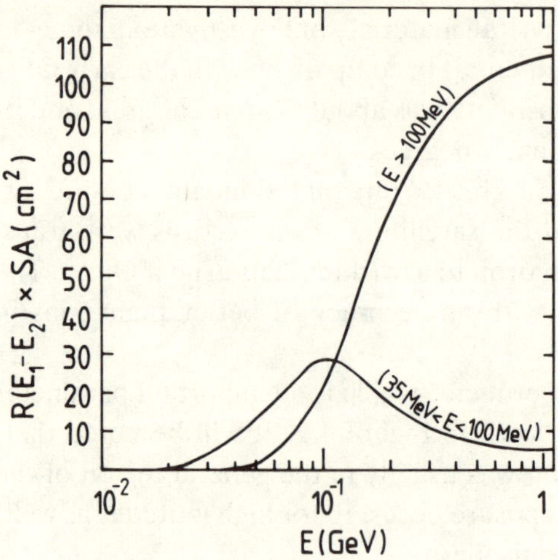

Figure 4.3. Energy response of the SAS II experiment. The ordinate is the product of the probability of finding a photon of energy E in the range E_1–E_2, the exposure factor and the effective detection area. (Lebrun and Paul 1983.)

gave 'response functions' for the two energy bands 35–100 and greater than 100 MeV as shown in Figure 4.3.

Of crucial importance in all branches of astronomy is the angular resolution of the telescope, and it is because this is so poor in γ-ray astronomy that there is so much argument about the interpretation of γ-ray data. The angular resolution of the satellite experiments is poor because of the nature of the pair production process and the scattering of the secondary electrons in the spark chamber plates. Figure 4.4 shows the angular resolution as a function of E_γ for SAS II.

4.3.2 A summary of the SAS II results

Figures 4.5 and 4.6 show the longitude and latitude dependences for γ-ray energies above 100 MeV. The main features evident are:

(i) A general concentration of flux in the Galactic Plane ($|b| \leqslant 3°$).

(ii) A general concentration towards the Galactic Centre but extending over $\pm 40°$ in longitude or so, thus confirming the feature observed earlier by OSO III (see Section 4.2).

(iii) Intensity peaks corresponding to the positions of two known pulsars, the CRAB and VELA (the identification being made positive by the detection of pulsed emission at the radio pulsar periods). A strong peak at $l = 195°$, $b = +4°$ (later named 'Geminga') from an object which

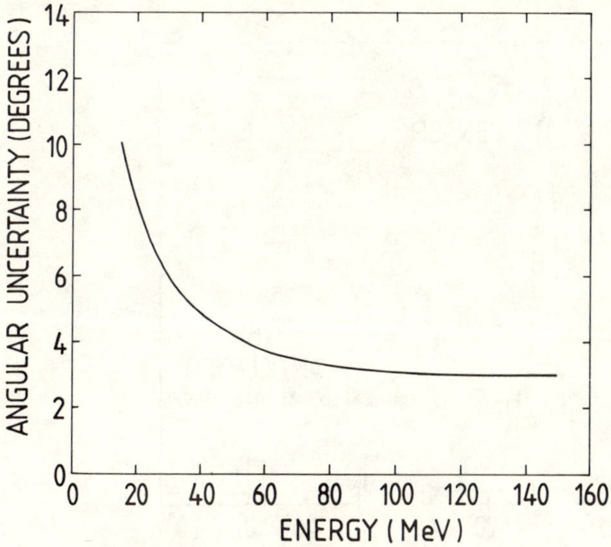

Figure 4.4. Three-dimensional angular uncertainty as a function of energy for SAS II (Kniffen *et al*. 1978).

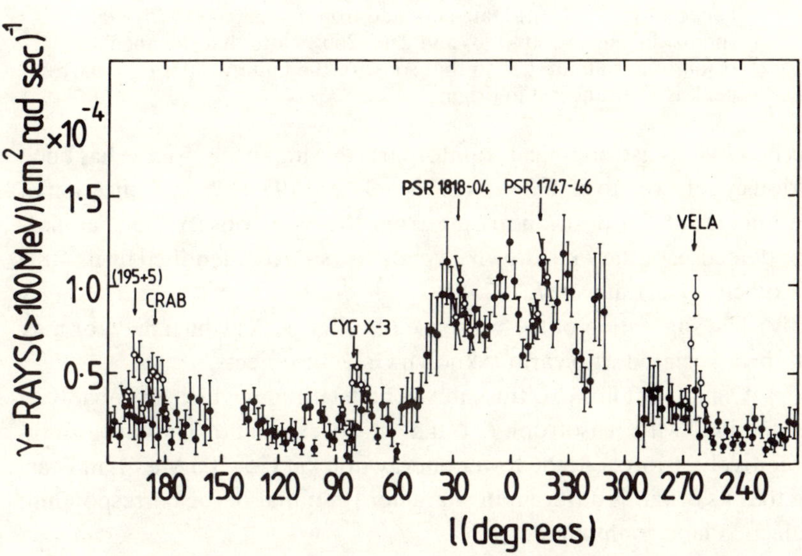

Figure 4.5. Distribution of γ-rays above 100 MeV along the Galactic Plane: SAS II (Thompson *et al*. 1976). The data refer to $-10° < b < 10°$. The diffuse background has been subtracted. Open circles include pulsed components. The two pulsars PSR 1818-04 and 1747-46 are not now thought to be significant γ-ray emitters, but the pulsed components for the other SAS II sources are still accepted.

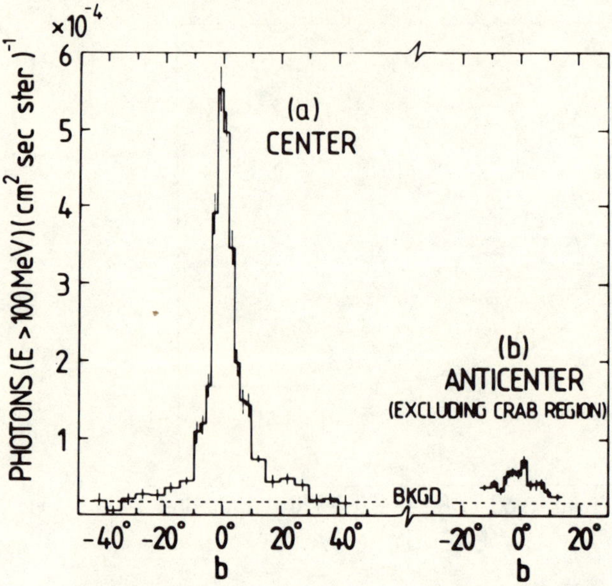

Figure 4.6. Latitude distribution of γ-rays above 100 MeV from the SAS II experiment (Fichtel *et al.* 1975). Dashed line: diffuse background level. (*a*) Data summed from l = 330°–30°. (*b*) Data summed from l = 90°–170° and 220°–260°. Note that the angular resolution is about ±3° so that some of the finite width of the narrow peak is instrumental in origin.

has no obvious astronomical counterpart (this important source has been variously referred to as 193 + 04, 195 + 04 and 195 + 05, an indication of the uncertainty in position arising from the γ-ray observations alone). The detection of the well known Cygnus X-3 source, identified by its 4.8 h periodicity (Hartman *et al.* 1976).

(iv) The suggestion of the presence of other peaks which may, or may not, be associated with various known Galactic objects.

(v) A small, but finite, diffuse flux at high latitudes which is probably to be identified with an isotropic extragalactic background. The background is relatively stronger in the lower energy bracket (35–100 MeV), indicating that its origin is different in character from that of the corresponding Galactic Plane component.

4.4 The COS B satellite

4.4.1 Experimental details

A European collaboration successfully designed, constructed and launched the COS B satellite carrying a detector rather similar to that

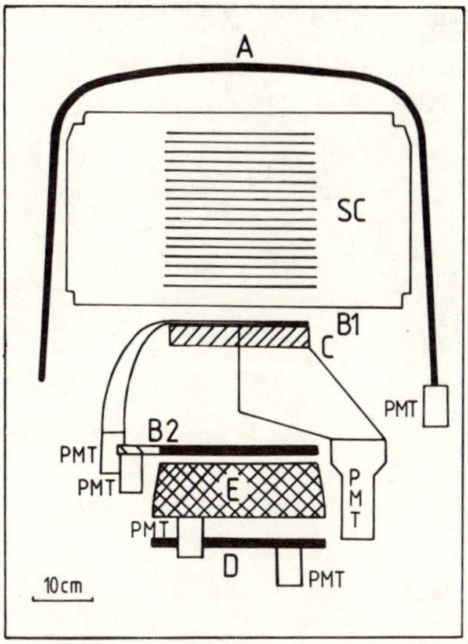

Figure 4.7. Sectional view of the COS B γ-ray detector (Bennett *et al.* 1976). A: anticoincidence counter; SC: spark chamber; B1, B2: scintillation counters, C: directional Cerenkov counter; E: energy calorimeter (caesium–iodide scintillator); D: scintillator to provide information on high energy events for which the absorption in the calorimeter is incomplete.

on SAS II. The launch occurred on August 9, 1975, and the instrument performed admirably for the next six and a half years until an already overdue gas shortage caused the mission to be finally terminated. The project was a triumph for international cooperation and marked a stepping stone for an increasing number of multi-nation collaborations.

The COS B detector is shown schematically in Figure 4.7; the similarity to the SAS II apparatus is immediately evident.

Apart from the difference in the lengths of exposure between COS B and SAS II the most important difference was the character of the satellite orbit: unlike the near circular orbit of SAS II, COS B was placed in a highly eccentric orbit. The philosophy was to increase the useful observation time by having only a small fraction of the sky occulted by the earth but the price paid was rather high: the apparatus was smaller than it could have been for a near-earth orbit and the flux of background cosmic ray particles was an order of magnitude higher. A consequence of the latter was a rather large background of secondary γ-rays produced in the detector and this has resulted in considerable difficulty in correcting the

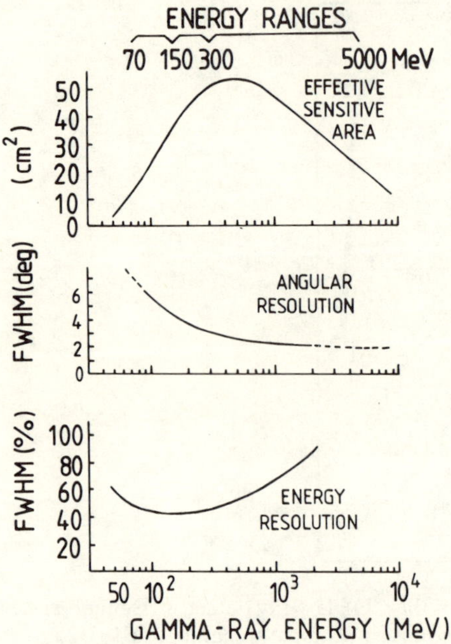

Figure 4.8. Characteristic parameters of the COS B experiment for γ-rays incident along the detector axis (Scarsi *et al*. 1981).

signal for background – an important feature for latitudes away from the Galactic Plane. An important consequence has been an inability to derive an estimate for the isotropic diffuse flux. Nevertheless, a large number of cosmic γ-rays were recorded ($\approx$100 000) allowing a number of important discoveries to be made.

Prelaunch calibration was performed, as with SAS II, using tagged-photon beams, the energy range involved being 20 MeV–6 GeV. The important sensitivity versus γ-ray angle relative to the spark chamber pointing direction was also determined. Selection criteria were devised relating to the pulse height thresholds for the counters of the triggering telescope, the anticoincidence counter and the energy calorimeter. Finally, detailed procedures were devised for the scientists who reduced and analysed the data. The relevant instrumental characteristics, effective sensitive area and angular and energy resolutions, are shown in Figure 4.8.

The angular resolution of the detector is of particular importance for later analyses and more description is needed. For a Gaussian distribution of projected angle, θ_x, θ_y, the probability can be written as

$$P(\theta_x, \theta_y) = C' \exp - [(\theta_x^2 + \theta_y^2)/2\sigma^2] \quad (4.1)$$

where C' is a normalisation constant and σ_ψ, the half angle of a cone such that 68% of the apparent directions are within this angle of the true direction, is given by $\sigma_\psi = 1.56\sigma$.

In fact the actual angular distributions are not Gaussian but have extended tails, although these are reduced somewhat in the editing of the records. Following Hermsen (1980) a good approximation is

$$P(\theta_x, \theta_y) = C' \exp - [(\theta_x^2 + \theta_y^2)/\theta_0^2]^c \quad (4.2)$$

Typical values for θ_0 and c are, for the energies indicated: 4.7°, 0.8 (75 MeV); 3.0°, 0.7 (150 MeV); and 1.25°, 0.5 (300 MeV).

4.4.2 A summary of the COS B results

A number of γ-ray maps have been published by the COS B group and a copy of an important set, which has been extensively used in analyses, is reproduced in Figure 4.9 (Figure 4.10 shows the associated intensity cuts). Inspection of the maps shows the presence of many 'hot spots' of γ-ray emission as well as peaks at the SAS II source positions: the CRAB, VELA and '195 + 5'. It is the detection of these candidate sources (hot spots) which has probably been the most important feature of the COS B results.

The longer exposure time for COS B has provided better statistics and, in turn, allowed observations to be made to higher energies than was possible with SAS II. For most of the sky the intensities are quoted for the three energy bands already referred to and for the two pulsars, which were the subject of extended observing, data have been recorded to 1 GeV. The selection of the highest energy band (300–5000 MeV) has turned out to be very important because the generally accepted view is that the bulk of these γ-rays are generated by way of π^0-decay, the pions themselves having been produced by the interactions of the cosmic ray protons and nuclei with the ISM. The search for the origin of cosmic ray protons and nuclei, as distinct from electrons, is bound up with these energetic γ-rays.

4.5 Gamma-ray sources

4.5.1 Introductory remarks

The priority to be attached here to the various satellite results reflects the preference of the authors but there can be no doubt that the 'γ-ray sources' loom large in any such discussion, and this is the topic with

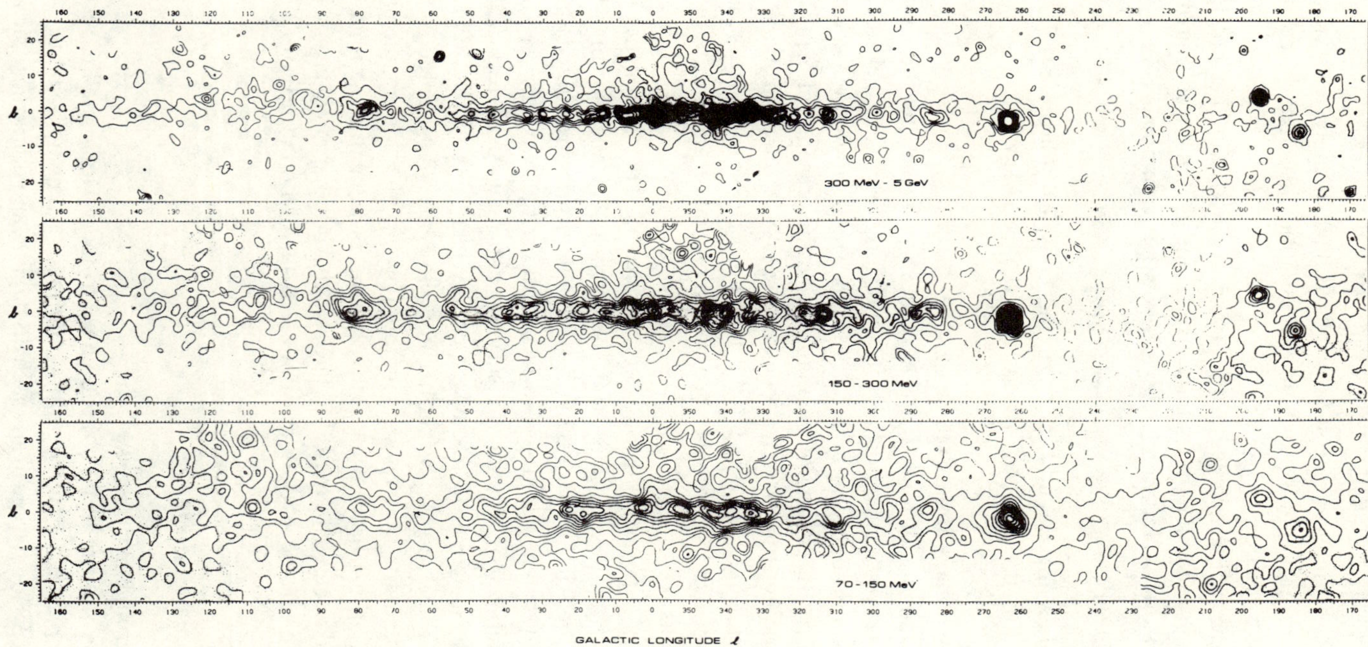

Figure 4.9. Contours of COS B γ-ray intensity from the work of Mayer-Hasselwander *et al.* (1982). The contour levels are indicated as multiples of 5×10^{-5} cm^{-2} s^{-1} sr^{-1} for 70–100 MeV, 3×10^{-5} for 150–300 MeV and 4×10^{-5} for 300–5000 MeV.

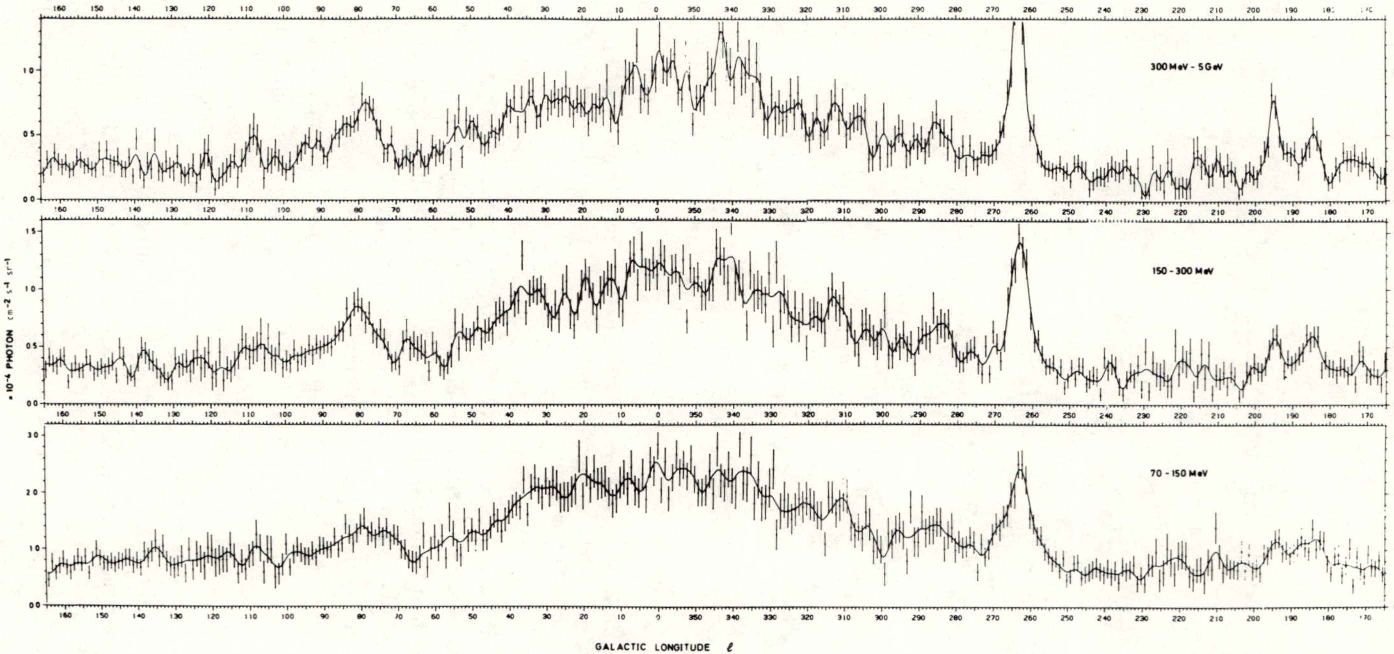

Figure 4.10. Longitude profiles of COS B γ-ray intensity averaged over $\pm 10°$ in latitude. Solid line: surface fitted to the data. (Mayer-Hasselwander *et al*. 1982.)

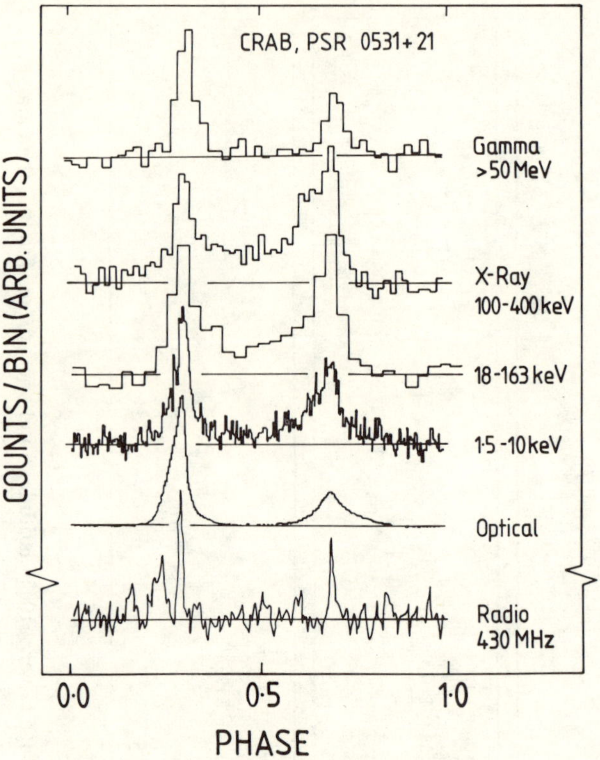

Figure 4.11. Light curves for the CRAB pulsar at various wavelengths. The γ-ray data are from a summation of five COS B observation periods (Wills *et al.* 1980).

which we will start. Attention will be devoted in turn to the pulsars, other identified sources and finally to the unidentified objects and the thorny question of their nature and the related question of the magnitude of the flux expected to arise from unresolved weak sources.

There have been a number of reviews of 'γ-ray sources' already; Fazio (1973) has covered the early observations and, most recently, a detailed survey of Galactic sources has been given by Bignami and Hermsen (1983).

4.5.2 Gamma-ray pulsars

As mentioned already, the first definite observation of γ-rays from pulsars came from the SAS II experiment with the detection of the CRAB (PSR 0531 + 21) and VELA (PSR 0833 − 45) (Thompson *et al.* 1975). Possible detections were also claimed for two old pulsars: PSR

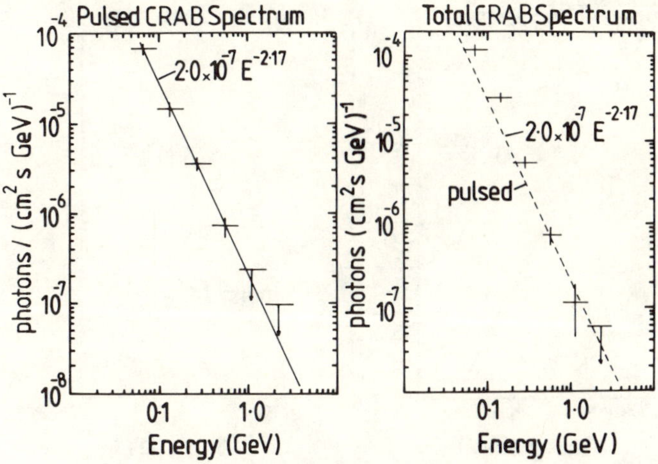

Figure 4.12. Differential photon spectrum for the CRAB γ-ray source (2CG 184−05). The line gives the fit to the pulsed component. It is apparent that the pulsed fraction increases with energy (after Hermsen 1981).

1747 − 46 and PSR 1818 − 04 (Ogelman *et al.* 1976). COS B extended the observations of the major sources and found new features but, after some tentative sightings of other pulsars, the view now is that only PSR 0531 + 21 and PSR 0833 − 45 have been definitely detected.

Starting with the CRAB, observations have been made essentially over the whole of the electromagnetic spectrum, from radio waves to γ-rays. Although results are described here up to 1000 MeV, later we will examine still higher energies – up to 10^{16} eV. Figure 4.11, from Bignami and Hermsen's review, summarises measurements of the pulsar 'light curve' at various wavelengths. It is interesting to note that although the intrinsic widths of the two pulses are similar at all wavelengths (except radio) there are a number of differences:

(i) the ratio of intensity of the interpulse to the main pulse is a function of wavelength (and there is COS B evidence that this ratio is variable above 50 MeV on a time scale of years, Wills 1981);

(ii) the valley between the pulses has a wavelength dependent magnitude.

A notable feature of the energy spectrum at X-ray and γ-ray energies is the increasing fraction of the pulsed component as one proceeds to higher energies; Figure 4.12 shows this feature.

An interesting aspect that follows from the detected spectral shape is that the maximum power of the pulsar is radiated in the γ-ray region,

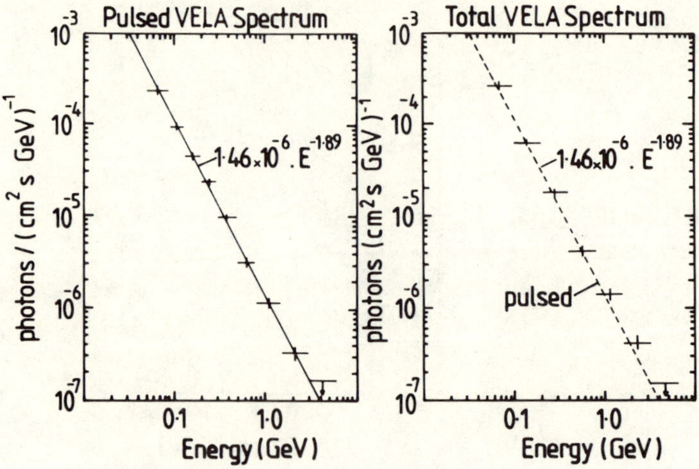

Figure 4.13. Differential photon spectrum for the VELA γ-ray source (2CG 263−02). The line gives the fit to the pulsed component. Most of the emission is seen to be pulsed at all energies (after Hermsen 1981).

specifically near 10 MeV. Buccheri (1980) has used the known distance of the pulsar, together with the assumption that emission is in two cones, to derive the rate of emission of energy in the γ-ray region: 2×10^{35} erg s^{-1} above 50 MeV. This emission is several orders of magnitude greater than that in the radio region and corresponds to about 0.04 % of the energy loss corresponding to its spin down rate.

Moving to VELA, this is the strongest γ-ray source in the sky at the energies in question (a flux above 100 MeV of 13×10^{-6} cm^{-2} s^{-1}). The energy spectrum (Figure 4.13) is interesting in that the differential exponent ($\gamma = 1.89$) is small and the pulsed fraction is very large indeed; Lichti *et al.* (1980) estimate that at least 90 % of the flux is pulsed above 50 MeV. Assuming isotropic emission, the luminosity above 50 MeV is $\approx 4 \times 10^{34}$ erg s^{-1}, corresponding to about 0.3 % of the spin down energy loss rate, i.e. a factor 10 higher than that of the CRAB. There are other differences from the CRAB, too; for example, the light curves for VELA are very different in the various energy regions: the phase difference between the main pulse and the interpulse differs between γ-rays and the optical region, and for the radio signal there appears to be no interpulse at all. Another remarkable difference is the lack of a detectable X-ray signal from VELA (Knight 1981). As will be described later, there have been detections of VELA at much higher photon energies ($>5 \times 10^{11}$ eV), but an increased spectral exponent ($\gamma \approx 2.65$) is needed at these energies in order to join the higher and lower energy regions.

A number of models have been proposed to explain the γ-ray emission from the CRAB and VELA and for pulsars in general. The inverse Compton mechanism is a strong contender (Apparao and Hoffman 1970, Cavaliere *et al.* 1970, and others), the idea here being that energetic electrons from the pulsar generate synchrotron photons, primarily in the X-ray region, and these X-rays are boosted in energy into the γ-ray domain by inverse Compton collisions with the electrons. Schlickeiser (1981) has examined this possibility in detail, paying regard to the accurate formulation of the interaction cross section. He concludes that the energy spectra of γ-rays from the CRAB and VELA can best be understood if the Compton interactions occur near the velocity of light cylinder. No doubt the differences in light curves (Figure 4.11) for the various parts of the electromagnetic spectrum are related to the fact that the quanta originate in different regions (e.g. the radio emission probably comes from near the polar caps); there is plenty of scope here for further detailed examination.

4.5.3 The 2CG gamma-ray source catalogue

Before discussing further individual sources it will be useful to consider the overall situation with regard to γ-ray sources. The most comprehensive work along these lines has been carried out by the COS B group, who have produced the '2CG catalogue' (Hermsen 1981) reproduced in Table 4.1. Figure 4.14 shows the positions and fluxes of these sources on a longitude–latitude (l, b) plot.

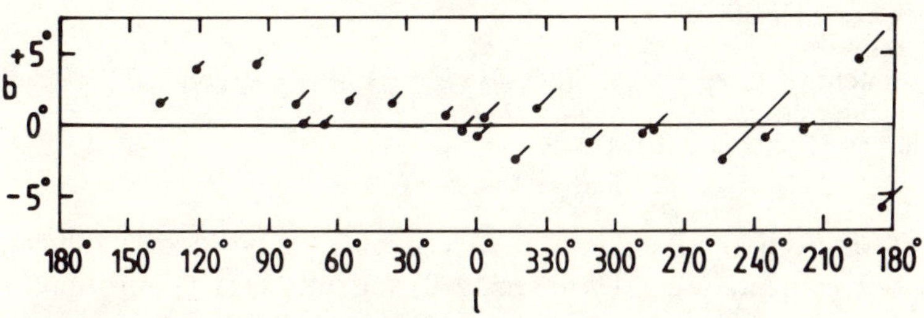

Figure 4.14. 2CG sources from the COS B catalogue of Hermsen (1980, 1981). The dots mark the estimated source positions (typical error radius 1°) and the inclined lines have lengths proportional to the quoted fluxes. The smallest flux shown is 1.0×10^{-6} cm^{-2} s^{-1} for $E_\gamma > 100$ MeV; the largest is for the VELA pulsar: 13.2×10^{-6} cm^{-2} s^{-1}.

Table 4.1. *The 2CG catalogue of gamma-ray sources (Hermsen 1981)*

Source name	Observation periods	Significance (σ_0')	Energy threshold (MeV)	Position l (degrees)	b	Error radius (degrees)
2CG006−00	2, 18, 25	10.2	300	6.7	−0.5	1.0
2CG010−31	30	5.7*	100	10.5	−31.5	1.5
2CG013+00	2, 8, 18, 25	5.3	300	13.7	0.6	1.0
2CG036+01	9, 25, 26	4.9	300	36.5	1.5	1.0
2CG054+01	9, 25, 26	5.3	100	54.2	1.7	1.0
2CG065+00	4, 9, 22, 26	5.5	100	65.7	0.0	0.8
2CG075+00	4, 16, 22, 26, 36	5.8	100	75.0	0.0	1.0
2CG078+01	4, 16, 22, 26, 36	11.9	100	78.0	1.5	1.0
2CG095+04	4, 16, 22	4.9	150	95.5	4.2	1.5
2CG121+04	11, 16, 28	4.9	100	121.0	4.0	1.0
2CG135+01	11, 16, 28	4.9	100	135.0	1.5	1.0
2CG184−05	1, 14, 17, 29	20.6	100	184.5	−5.8	0.4
2CG195+04	1, 14, 29	27.1	100	195.1	4.5	0.4
2CG218−00	14, 18, 21	6.2	100	218.5	−0.5	1.3
2CG235−01	19, 21	5.0	150	235.5	−1.0	1.5
2CG263−02	3, 5, 12, 21	35.7	100	263.6	−2.5	0.3
2CG284−00	5	6.5	100	284.3	−0.5	1.0
2CG288−00	5	4.8	100	288.3	−0.7	1.3
2CG289+64	10, 32	6.5*	100	289.3	64.6	0.8
2CG311−01	5, 7	5.6	150	311.3	−1.3	1.0
2CG333+01	7, 13, 24	5.4	300	333.5	1.0	1.0
2CG342−02	2, 7, 13, 18, 24	8.9	300	342.9	−2.5	1.0
2CG353+16	2, 13, 18, 24	5.1*	100	353.3	16.0	1.5
2CG356+00	13	5.3	300	356.5	0.3	1.0
2CG359−00	2, 18, 24	6.3	300	359.5	−0.7	2.0

* expressed in units of σ_0.
** assuming E^{-2} spectra.
*** intensity ($E > 300$ MeV)/intensity ($E > 100$ MeV), assuming E^{-2} spectra calculating both intensities.
The error radius is at the 90 % level.

The selection criteria used in determining the positions and fluxes are well documented (e.g. Hermsen 1980) and only a brief discussion will be given here.

The so-called 'cross correlation technique' was used to search for statistically significant excesses above the adjacent background values, a technique that involves knowledge of the detector point spread function (PSF). Ideally, one should use only data for which the PSF is narrowest, namely the highest energies, but the small number of γ-rays recorded at

Flux** $E > 100$ MeV (10^{-6} photons $cm^{-2} s^{-1}$)	Spectral*** parameter	Comments: CG source	Identification	Other
2.4	0.39 ± 0.08			
1.2				
1.0	0.68 ± 0.14			
1.9	0.27 ± 0.07			
1.3	0.20 ± 0.09			
1.2	0.24 ± 0.09	CG064+00		
1.3		CG075−00 }		could be an extended feature
2.5		CG078+01 }		
1.1				
1.0	0.43 ± 0.12	CG121+03		
1.0	0.31 ± 0.10	CG135+01	CTO236+610?	
3.7	0.18 ± 0.04	CG185−05	CRAB pulsar	
4.8	0.33 ± 0.04	CG195+04		
1.0	0.20 ± 0.08			
1.0				
13.2	0.36 ± 0.02	CG263−02	VELA pulsar	
2.7		}		could be an extended feature
1.6		}		
0.6	0.15 ± 0.07	CG291+65	3C273	
2.1		CG312−01		
3.8		CG333+00		
2.0	0.36 ± 0.09			
1.1	0.24 ± 0.09		ρ Oph. dark cloud?	
2.6	0.46 ± 0.12			may be variable
1.8				

high photon energy dictates use of a compromise value; in the COS B case 100 MeV was chosen for most of the analysis. The basic requirement was that the peaks derived by adopting the cross correlation technique should be consistent with 'point' sources and that the excess fluxes should be at a high level of significance, symbolised by a probability of being spurious of $\lesssim 10^{-6}$. It was this latter condition that gave the effective threshold of source flux of 1.0×10^{-6} $cm^{-2} s^{-1}$ above 100 MeV apparent in Table 4.1, except for the high latitude source 2CG 289 + 64 (identified with the quasar 3C273) which was the subject of special treatment.

The uncertainty in position is clearly an important limitation in attempts to identify γ-ray sources with objects visible at other wavelengths. An order of magnitude estimate of what this uncertainty should be can be made using the data on angular resolution given earlier in Figure 4.8. There the FWHM is seen to be $\approx 4°$ for a mean energy corresponding to γ-rays above 100 MeV; the corresponding error circle at the 90 % level has a radius of $\approx 4.5°$. Now a typical 'source' with flux near $1.0 \times 10^{-6}\ \text{cm}^{-2}\ \text{s}^{-1}$ is due to the detection of about 100 γ-rays so that we might expect an error circle of radius $\approx 4.5°/100^{1/2} = 0.45°$. In fact, because of the presence of the background against which the source is viewed, the resolution might be expected to be degraded by a factor of, typically, 2, i.e. to $\approx 1°$. Inspection of Table 4.1 shows that the radii of the positional uncertainty circles are indeed in this region.

From some points of view the 2CG catalogue represents the most important result from the COS B experiment for γ-ray astronomy in this energy bracket and it has provoked considerable discussion and argument, particularly with regard to the question of whether those sources which have not been identified with 'discrete' objects visible in other wavelengths are genuine or not (see, for example, the useful summary by Hermsen and Bloemen 1982). This topic will be examined in depth later but some discussion will also be given in this section.

The first question to be considered is whether or not there is support for there being genuine γ-ray peaks at the COS B 'source' positions, whatever the origin of the peaks; in other words, can we discount technical imperfections and statistical fluctuations? The answer is that most (and perhaps all) of the peaks seem to be genuine, this conclusion following from the SAS II observations, both by way of the identification of some of the obvious sources, e.g. CRAB, VELA and Geminga, by the SAS II workers themselves (see Section 4.3.2), and by way of a more recent analysis by Houston and Wolfendale (1983). Table 4.2 gives a summary of the last mentioned analysis; in it, the cross correlation technique of Hermsen, as used for COS B data, was applied to the basic SAS II data of Fichtel *et al.* (1978). It will be noted that there is strong support for the presence of genuine peaks in the positions of a number of the 2CG sources and there is no objection to most of the others being genuine in the same sense, too.

The argument centres rather on the cause of the peaks. The alternative view to their explanation as discrete sources, such as previously unidentified pulsars, is that they are due to cosmic ray irradiation of the lumpy interstellar medium. It is well known that there are many dense clouds of

Table 4.2. *Analysis of SAS II data by Houston and Wolfendale (1983)*

(a) SAS II candidate source catalogue for $E_\gamma > 100$ MeV with significance, measured as C_s/σ_0', greater than 3.4. Flux F_γ, m measured in flux units at the position of maximum significance. F_γ, s: flux measured at 2CG position. F_γ, COS B denotes the actual flux measured by the COS B workers and given in the 2CG catalogue.

2CG	l (degrees)	b (degrees)	Significance	F_γ, m	F_γ, s	F_γ, COS B
359−00	0.0	−1.2	4.5	2.3 ± 0.5	2.2	1.8
078+01	78.2	1.2	6.0	3.2 ± 1.2	2.9	2.5
	125.3	−0.4	5.0	3.8 ± 1.3		
184−05	185.3	−3.6	15.1	3.6 ± 0.8	2.0	3.7
195+04	196.0	5.2	5.0	4.9 ± 1.3	3.8	4.8
263−02	264.1	−2.8	17.1	10.7 ± 1.5	9.0	13.2
	25.0	8.4	3.5	1.4 ± 0.6		
	105.0	1.2	3.7	2.1 ± 0.9		
	114.9	−4.4	3.4	1.7 ± 0.7		
	312.0	7.6	3.7	2.9 ± 1.4		
311−01	314.8	−0.4	3.9	5.6 ± 2.0	0.0	2.1

(b) Data on SAS II peaks ($E_\gamma > 100$ MeV) within 3.6° of those 2CG sources not identified in Table 4.2(*a*). Significance measured as C_s/σ_0'. *W* indicates that the profile had the correct width. Flux F_γ, m measured in flux units at the point of maximum significance. F_γ, COS B is the flux measured by the COS B workers. Blank line indicates no peak found. None of the significances is above our lower limit for acceptance (3.4), but it will be seen that the actual values found are not too far away (>2.0). There is thus some support for nine more 2CG sources and rather more support for the seven which have correct *W*-values.

2CG	l (degrees)	b (degrees)	Significance	Width	F_γ, m	F_γ, COS B
006−00	8.0	−0.4	2.5		1.9 ± 0.9	2.4
013+00					(<0.7)	1.0
036+01	35.3	0.4	3.9	*W*	2.0 ± 1.0	1.9
054+01	54.4	1.2	2.3	*W*	1.9 ± 1.1	1.3
065+00					(<1.3)	1.2
075+00	74.5	−1.2	2.5	*W*	2.4 ± 1.3	1.3
095+04					(<1.5)	1.1
121+04	118.1	6.0	3.1	*W*	1.8 ± 0.9	1.0
135+01	135.0	2.0	3.2	*W*	2.7 ± 1.7	1.0
218−00	216.7	−2.0	2.4		1.5 ± 0.8	1.0
235−01	235.0	−0.4	1.7		1.1 ± 0.5	1.0
284−00	283.0	−2.0	2.1	*W*	2.2 ± 1.2	2.7
288−00					(<2)	1.6*
333+01	334.0	0.4	2.3	*W*	3.9 ± 1.7	3.8
342−02	343.0	−1.2	1.5	*W*	1.9 ± 1.3	2.0
356+00					(<2)	2.6*

* Note 2CG 356+00 'may be variable' and 2CG 284−00 and 288−00 'may be extended'.

molecular gas in the Galaxy, and some of these, on being penetrated by cosmic rays of the same intensity as those locally, might give rise to γ-ray peaks which would pass the COS B selection criteria. When we include the possibility, perhaps even the 'probability', of cosmic ray particle sources (SN, OB star associations, etc.) being present in these 'giant' molecular clouds, thereby increasing the cosmic ray intensity in the clouds relative to the local value, the potential of irradiated clouds for mimicking γ-ray sources is seen to be rather large. This topic of pseudo-sources is returned to after attention is given to other γ-ray sources which are not the subject of as much dissent.

4.5.4 Cygnus X-3

At much higher γ-ray energies (10^{11}–10^{16} eV), the well known X-ray source Cygnus X-3 appears to be a dramatic emitter (see Sections 5.2 and 5.3 for a discussion) and it is appropriate to examine its γ-ray emission in the present energy bracket in some detail. Its well known characteristic 4.8 h period in the X-ray region (first seen by Parsignault *et al.* 1972) led many γ-ray observers to search for this period in their own energy bands. Balloon observations from 1972 onwards claimed high fluxes, at least until 1977 or so (e.g. Galper *et al.* 1977). SAS II claimed a significant signal (Lamb *et al.* 1977), the pulsed flux above 100 MeV being $(4.4 \pm 1.1) \times 10^{-6}$ cm^{-2} s^{-1} and that above 35 MeV $(10.9 \pm 3.1) \times 10^{-6}$ cm^{-2} s^{-1}. These fluxes, although 'large' compared with the 2CG catalogue given in Table 4.1, are appreciably smaller than those of Galper *et al.* (1977), leading to the likelihood that the object is waning. Support for this contention comes from the COS B observations – no pulsed emission was detected in 1975 (Bennett *et al.* 1977) nor in later years up to 1980 (Swannenburg *et al.* 1981).

Figure 4.15 gives a summary by Rana *et al.* (1984) of the emission from Cygnus X-3 over the last decade or so – there is seen to be rather good evidence for a gradual fall-off of intensity. Further discussion of this remarkable object, which is almost certainly a binary system involving an energy outflow of about 10^{38} erg s^{-1}, is deferred to Sections 5.2, 5.3 and 5.4.

4.5.5 Geminga

It is a remarkable fact that only recently has a satisfactory explanation been given for the nature of this, the second strongest γ-ray source at the energies in question. The strength of the source, its lack of significant extension beyond the PSF of the COS B instrument, and the

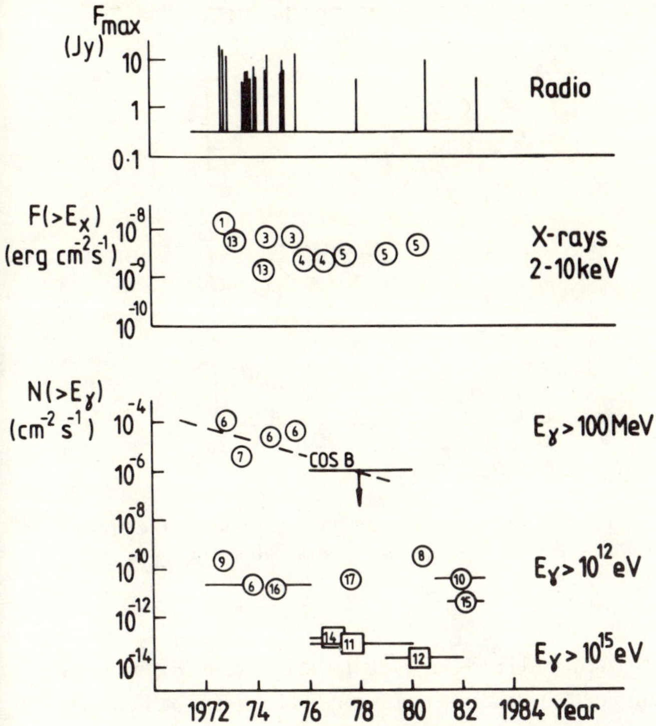

Figure 4.15. Summary of observations for Cygnus X-3 (after Rana *et al*. 1984). The ringed numbers refer to the various experiments.

absence of known molecular clouds in that direction, lead to a consensus interpretation as a 'discrete source'. So important has this source been regarded that five observing periods were devoted to it in the COS B project and nearly 1000 γ-rays were detected from it.

Inspection of optical, radio and X-ray records have revealed a number of possible candidates in the (0.4° radius) error box of Geminga but most had a number of unsatisfactory features (see Schlickeiser 1981 for a useful summary and comments on the likely production mechanism). The best identification has come from a very recent detailed search by Bignami and associates (described by Bignami 1984) and this result will be described in some detail.

With the advent of the Einstein X-ray Observatory, having three orders of magnitude higher sensitivity, Bignami *et al*. (1984) were able to make a careful examination of the Geminga error box. The result was the detection, at a level of $\approx 10^{-12}$ erg cm^{-2} s^{-1}, above a few keV, of a rather

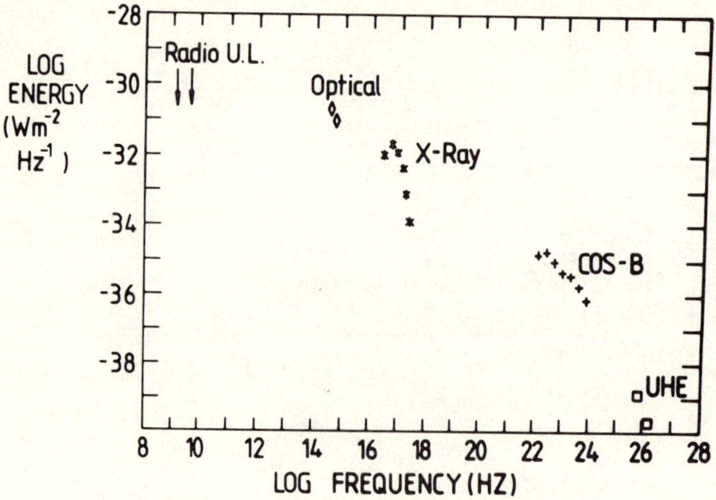

Figure 4.16. Summary of observations on the γ-ray source 'Geminga' (2CG 195+04) from the work of Bignami (1984). This object was seen first in γ-rays.

peculiar X-ray source (denoted IE 0630 + 178 in the Einstein Observatory catalogue). The main peculiarity was the absence of a star in the Palomar Observatory sky survey plates. A special optical search was therefore instituted to look for a fainter star than the Palomar limit. The search was successful (Caraveo *et al.* 1984) in that a very faint object with magnitude $M_r = 20.3$ was detected in the 50 square arc second error box occupied by the X-ray source. The spectral shape of the candidate, exemplified by $\approx 10^{-15}$erg cm^{-2} s^{-1} in the visible, 10^{-12} erg cm^{-2} s^{-1} in X-rays and 10^{-9} erg cm^{-2} s^{-1} in γ-rays (above 100 MeV), makes it a strong contender. Figure 4.16 shows the whole energy range.

What seems to clinch the identification is the apparent time variability at γ-ray and X-ray wavelengths. The γ-ray situation can be considered first. Thompson *et al.* (1977) gave evidence for a 59 s periodicity in the emission from Geminga, based on the 121 γ-rays detected, and they made a case for a finite rate of change of period over the six-month observing period: $\dot{P} \approx 2 \times 10^{-9}$ s s^{-1}. Clearly the significance was not great but this was increased after the launch of COS B when Masnou *et al.* (1977) claimed confirmation of both P and $\dot{P}$ from an analysis of the early COS B results. Remarkably (as it subsequently turned out) in a later publication the COS B collaboration retracted its confirmation, despite three years of satellite operation, on the grounds of a previous underestimate of

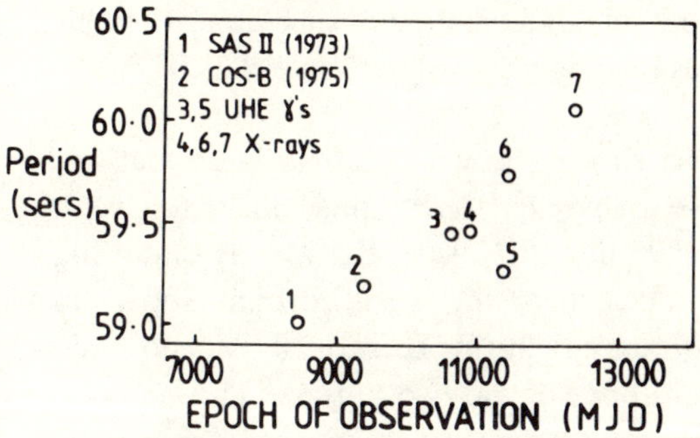

Figure 4.17. Period versus epoch for the Geminga γ-ray source (Bignami *et al.* 1984) (in fact, there is still argument about the significance of the SAS II and COS B results). A likely model for the source is that it is a binary pulsar, which is slowing down rapidly.

the statistical uncertainty. The whole of the COS B data are currently being analysed (Bignami *et al.* 1984) and, based on some three times the data used in the 'retraction' work, this problem should hopefully be solved rather soon.

We return to X-rays, however, for what appears to be an elucidation of the situation. Bignami *et al.* (1984) followed up the identification of the candidate X-ray source with the Einstein Observatory, with an observing run on EXOSAT (in 1983) and inspection of previous data with Einstein for 1981 and 1979. The results all now seem to fall into a pattern, as shown in Figure 4.17, with the period increasing systematically and smoothly with time. The evidence here for the identification of Geminga with the low luminosity X-ray star is strong; confirmation from a careful examination of the whole COS B data base will surely follow.

Some general points can be made before turning to specific models. The age of the object is presumably short, $\approx 10^3$ y (for $\dot{P} = 2 \times 10^{-9}$ s s^{-1}, $P/\dot{P} \approx 3 \times 10^{10}$ s). Its distance is unclear but the lack of obvious absorption features in either X-rays or the visible implies that it is closer than about 200 pc (Bignami *et al.* 1984). The very flat energy spectrum (Figure 4.16), and particularly the lack of observed radio emission, indicate a near unique object.

Turning to specific models, a number have been put forward. A young pulsar would appear at first sight to be favoured but the lack of the associated SNR and the absence of the characteristic radio signal present difficulties. The most recent, and probably the best, is that due to Fabian and Nulsen (1984). Here, a binary system is proposed comprising two magnetised neutron stars; it having been hypothesised by Henrichs and

Van der Heuvel (1983) that the rate of formation of such binaries (which might have been thought of as rare) is $\approx 3 \times 10^{-4}$ y^{-1}, i.e. 1 % of the rate of formation of single pulsars.

The difficulty of the rapid slowing down of an isolated pulsar is circumvented and it is possible to invoke a specific mechanism to give the observed γ-rays. Following Fabian and Nulsen, the idea is that a high potential difference is set up across the small orbital separation ($\approx 10^{10}$ cm), this potential difference being in excess of 10^{12} V, and arising between certain magnetic field lines of the two pulsars. The γ-rays, in turn, come from the 10^{12}–10^{13} eV electrons by way of synchrotron and/or curvature radiation.

In the Fabian–Nulsen model, the observed X-ray and γ-ray periods, or rather frequencies, are the difference between the spin frequency of the primary neutron star (that with the larger magnetic moment) and the orbital period of the other neutron star. The detailed calculation requires Geminga to be at a distance of 11 pc, ignoring any beaming effects. An interesting feature is the prediction that the system is a strong source of gravitational radiation; specifically the loss rate in this mode is $\approx 3 \times 10^{36}$ erg s^{-1}, a value greater than that in electromagnetic radiation. Unfortunately, gravitational wave detectors of adequate sensitivity are not yet available to check this prediction.

Perhaps the biggest drawback of the model is the comparative unlikelihood of finding such an object nearby. With a birth rate of 3×10^{-4} y^{-1} and a lifetime (for the γ-ray mode) of 10^3 y we expect only about 0.33 anywhere in the Galaxy at any one time; the chance of one being within say 15 pc is very small indeed ($\approx 10^{-7}$). In the same vein, if the output of 10^{31} erg s^{-1} in γ-rays above 100 MeV were typical of a volume of radius 15 pc the Galactic emission from such sources alone would be $\approx 10^{38}$ erg s^{-1}, somewhat higher than the known total emission of γ-rays from the Galaxy. We must conclude that, whatever its precise nature, if it is as close as 15 pc then it is there by a rare chance.

4.5.6 The possibility of pseudosources

As has already been remarked, there has been considerable argument as to the nature of the unidentified COS B sources (see the 2CG catalogue in Table 4.1). The poor angular resolution of the detectors, coupled with the uneven γ-ray background, makes it inevitable that doubts arise.

Although the fact that the 2CG catalogue included sources whose statistical significance was shown by straightforward calculation to be very considerable, the crucial role played by the lumpy background – γ's from cosmic rays interacting with the interstellar gas, denoted 'CRI' ('cosmic ray irradiation of the ISM') – does not appear to have been realised until the work of Li and Wolfendale (1981). These authors used a statistical analysis based on the likely numbers of giant molecular clouds in the inner Galaxy ('direct' information by way of detailed CO maps were not then available) to claim that between 40 and 60 % of the 2CG sources could be explained solely on the basis of CRI. In fact, a new analysis of the masses of molecular clouds (to be described), which reduces them somewhat, means that the Li and Wolfendale estimate should be revised downwards a little (to 30–50 %).

The situation is epitomised by the 2CG source associated with Rho Ophiucus ('ρ Oph'). Initially, this was considered to be a discrete source but the accumulation of more γ-ray data by COS B led to the appreciation that it can be explained by CRI. Figure 4.18 shows the situation. It should be remarked that the object has not ceased to be of interest; in fact, it may be very important because of the information it gives about the magnitude of the cosmic ray intensity in that region. Although this topic is taken up again later it can be remarked that the cosmic ray intensity may be significantly higher there (Morfill *et al.* 1981, Issa and Wolfendale 1981*a*). Some other 2CG 'sources' may also be associated with molecular clouds which contain higher than average cosmic ray intensities (Section 4.6.4).

Perhaps the strongest evidence against accepting the totality of the 2CG sources as genuine comes from the work of Houston and Wolfendale (1984). These workers have drawn attention to the fact that, whatever the fraction of genuine sources in the data, some genuine sources will be lost because of the confusion caused by the lumpy background, and the confusion will be greatest in the inner Galaxy. Houston and Wolfendale argue that the hypothesis that the source catalogue is correct in its entirety can be checked by evaluating the loss, correcting for it and then deriving the total flux from all sources, resolved and unresolved.

A detailed study was made of the efficiency of detection of a source as a function of source strength, longitude and latitude. As expected, the efficiencies are often quite low. For example, even a comparatively strong source, of strength 1.3 f.u. (1 f.u. = $1 \times 10^{-6}\,\mathrm{cm^{-2}\,s^{-1}}$ with $E > 100$ MeV) has only a 20 % probability of being detected at a longitude $l = 30°$. A source of strength 1 f.u. – the most common source strength in the 2CG catalogue – has a detection probability of less than 10 % at $l = 30°$ and still only $\approx$50 % at $l = 120°$.

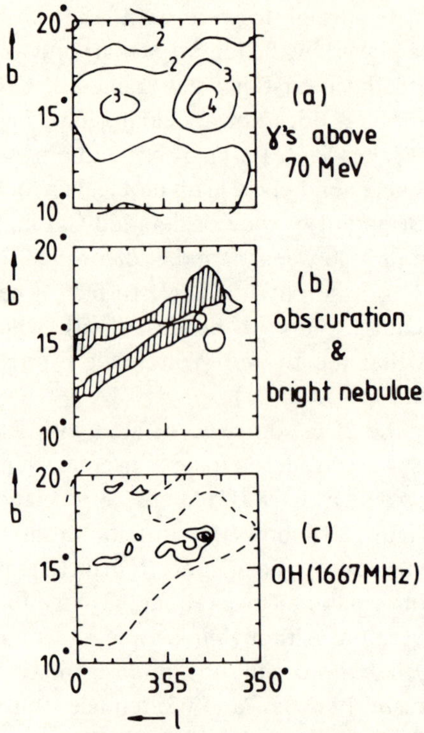

Figure 4.18. The Rho Ophiucus region. (*a*) γ-rays above 70 MeV from COS B (Mayer-Hasselwander *et al*. 1982). (*b*) An indication of cloud geometry from the *Atlas of Galactic Dark Nebulae* due to Khavtassi (1960). The shaded area relates to the heaviest obscuration recorded; 'bright nebulae' (unshaded) will also be targets for cosmic ray interaction. (*c*) Contours of CO emission from the 1667 MHz maps of Wouterloot (1981); the contour intervals are 0.06, 0.10, 0.14 K . . . It is apparent that the γ-ray emission roughly mirrors that of the gas.

The net result is that if all the 2CG sources are assumed genuine then the corrected source spectrum can be derived. Figure 4.19 illustrates the situation. Considering the inner Galaxy, the log N–log F_γ plot is seen to be very steep indeed, so steep in fact that for any reasonable choice of the radial (Galactocentric radius) distribution of sources, and their form must satisfy the corrected log N–log F_γ plot, the total integrated γ-ray flux, from both resolved and unresolved sources, is impossibly high. Quoting Houston and Wolfendale, with a radial source distribution of a form similar to that of SNR raised to the power p, i.e.

$$\rho(R) \propto \{\rho_{\mathrm{SNR}}(R)/\rho_{\mathrm{SNR}}(R=10)\}^p \quad (4.3)$$

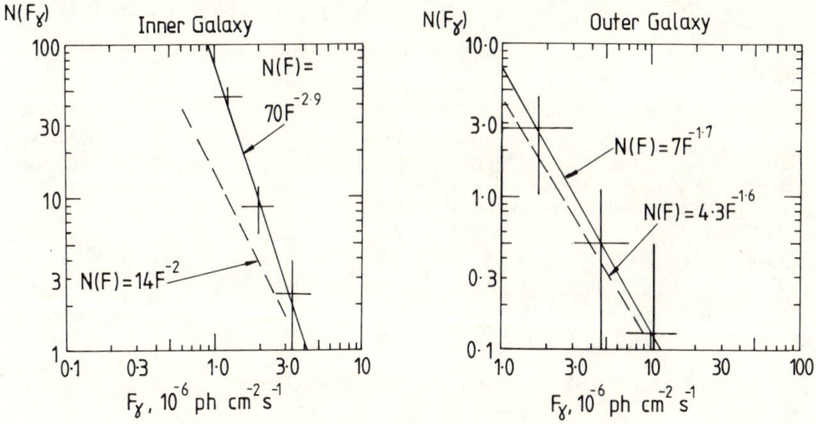

Figure 4.19. Differential distribution of source fluxes for COS B as corrected by Houston and Wolfendale (1984) to allow for loss by confusion (full lines). Their attempt at rejecting likely pseudo-sources due to cosmic ray-irradiated gas gives the 'genuine source' distributions shown by the dotted line.

where R is in kpc, $p = 3$ is needed and the total integrated flux above 100 MeV is 430 f.u. Now the total observed flux is only 230 f.u. so that this is an impossible situation.

Moving to the question, again, of what fraction of the γ-ray sources are genuine and what fraction of the total flux is due to genuine sources, recourse has to be made to models for their distribution in the Galaxy. An SNR-type distribution (equation 4.3) with $p = 1$ and individual γ-ray luminosities of the order of those observed for the few identified sources, give a model which appears to be consistent with all the observations, including the statistical estimate of pseudosources made earlier. The corresponding fractions of pseudosources are 60 % for the inner Galaxy and 30 % for the outer Galaxy. The log N–log F_γ plots for genuine sources are also shown in Figure 4.19. The corresponding fraction of the Galactic flux due to unresolved discrete sources is ≈20 % for the inner Galaxy and ≈15 % for the outer Galaxy. (This topic is considered in more detail later.)

These ideas concerning the fraction of pseudosources have received a measure of confirmation very recently with the work of Pollock *et al.* (1984, reported by Bignami *et al.* 1984). 2CG sources in the range l: 0°–90° have been reinterpreted in the light of the CRI problem referred to above, and using the most recent CO data, with the result shown in Table 4.3. Insofar as the energy threshold differs from that in the 2CG catalogue

Table 4.3. *Reanalysis of 2CG sources in the first quadrant (Bignami et al. 1984)*

Out of seven sources, three now appear consistent with cosmic ray-irradiated gas and there is a new one too.

Name	*l* (degrees)	*b* (degrees)	Flux ($\times 10^{-7}$ cm^{-2} s^{-1})	No. of obs.	Confidence parameter
(*a*) Objects detected above 300 MeV in addition to the gas					
2CG078+01	78.1	2.3	4.8	7	52.8
2CG075+00	76.1	0.5	3.8	7	
2CG054+01	55.0	2.2	5.9	1	9.2 variable
083+03	83.0	3.2	5.3	1	8.6 variable
2CG013+00	13.9	0.2	4.9	4	7.3
(*b*) Objects above 300 MeV satisfactorily explained by the gas, with limits on the extra flux					
2CG036+01			3.7	5	0.6
2CG065+00			2.1	4	0.0
2CG095+04			2.4	4	1.2

(300 MeV compared to 100 MeV), a proper comparison with previous predictions of the number of pseudosources cannot be made – in fact, the numbers involved are so small that a precise comparison is not possible in any case. Nevertheless, the situation is heartening. Out of seven sources, three have moved into the category of being satisfactorily explained in terms of CRI, a number not inconsistent with the 30–50 % referred to above.

Surprisingly, perhaps, one of the 2CG sources (054 + 01) is now categorised as 'variable', as is a new one at 083 + 03 (Pollock *et al.* 1981).

4.6 The diffuse Galactic emission

4.6.1 General remarks

Even as early as the first satellite experiments, notwithstanding their crude angular resolution, attempts were made to interpret what appeared to be a diffuse Galactic γ-ray component in terms of cosmic ray interactions (see, for example, the work of Stecker 1969, Clark *et al.* 1968, Fichtel *et al.* 1972, and Strong *et al.* 1973). These attempts have continued, with increasing sophistication, and also, it is hoped, with increasing accuracy in the end product.

Implicit in most of the analyses is the assumption that the contribution to the observed γ-ray flux from unresolved discrete sources in the Galaxy is small. In addition to the arguments presented already mention should also be made of a number of indirect estimates of the contribution; without exception, they give values in the region of 25 % of the observed

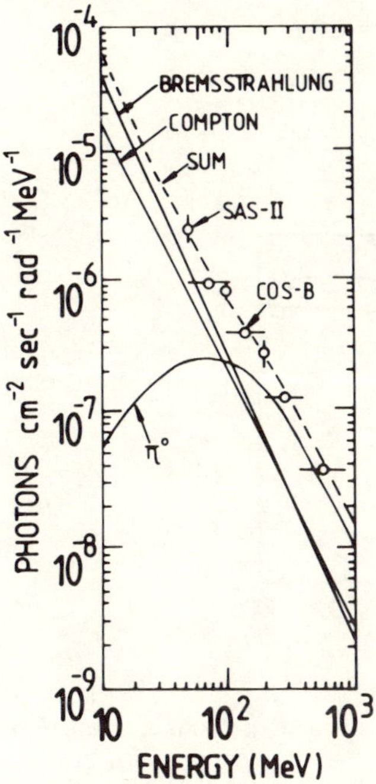

Figure 4.20. Energy spectrum of γ-rays from the general direction of the Galactic Centre. The calculated spectra are from Fichtel (1983) and represent a possible combination of the various components (the actual division is difficult to determine and is subject to argument). The experimental intensities are from Mayer-Hasselwander *et al.* (1982) for COS B and Fichtel (1983) for SAS II.

flux (e.g. Protheroe *et al.* 1979). A particularly strong suggestion that the contribution is indeed small is the fact that, along the Galactic Plane, the γ-ray intensity correlates well with the synchrotron radio emission (Haslam *et al.* 1981). The latter is accepted to be almost entirely of a diffuse nature and due to the spiralling of GeV electrons in the tangled Galactic magnetic field. Particularly safe from unresolved sources is the Galactic Anti-Centre (AC) direction ($l = 90°–270°$) and all longitudes for $|b| > 10°$, where most discrete sources are expected to be readily recognisable.

When examining the correlation of the diffuse γ-ray intensity with the column densities of gas in the ISM, to deduce the corresponding cosmic ray intensity, it is usually important to distinguish between the two major cosmic ray components, electrons and 'nuclei' (the latter are primarily protons – in fact, the nuclear component will often be referred to just as 'protons', although in practice all the nuclear components are included in the calculations). Figure 4.20 (from Fichtel 1983) gives an estimate of the

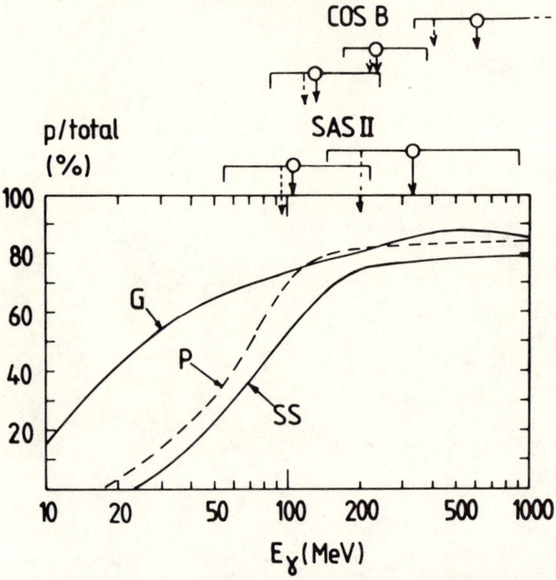

Figure 4.21. The fraction of the γ-ray flux due to protons as a function of γ-ray energy (Bhat *et al*. 1984*a*). The median energies for the energy bands for the SAS II and COS B measurements are indicated, as are the limits such that 67 % of the signal lies within. Dotted vertical lines: energies at which one should consult the graphs in order to find the appropriate proton fraction. G, P, SS: different authors' predictions (G: Goned 1981; P: Poon 1983, both for the 'local spectrum'; SS: Sacher and Schonfelder 1984, the 'anti-centre spectrum'; 116°–136°). It will be noted that protons are dominant for the higher band for SAS II and the two upper bands for COS B.

constituent γ-ray spectra for the general Galactic Centre (GC) direction and demarcates the regions where the proton (π^0) and electron (bremsstrahlung + Compton) components are likely to be important. An alternative manner of presentation, and one which indicates some of the disparity amongst the various predictions, has been given by Bhat *et al.* (1984*a*) and is shown in Figure 4.21. The fractions of the important proton component for the various energy ranges adopted in the SAS II and COS B experiments are indicated in this figure.

4.6.2 Gas in the Galaxy

Hydrogen is the dominant gas component in the Galaxy, with helium present to the extent of about 36 % by mass, and heavier nuclei contributing a further 2 %. Interstellar grains are probably responsible for about

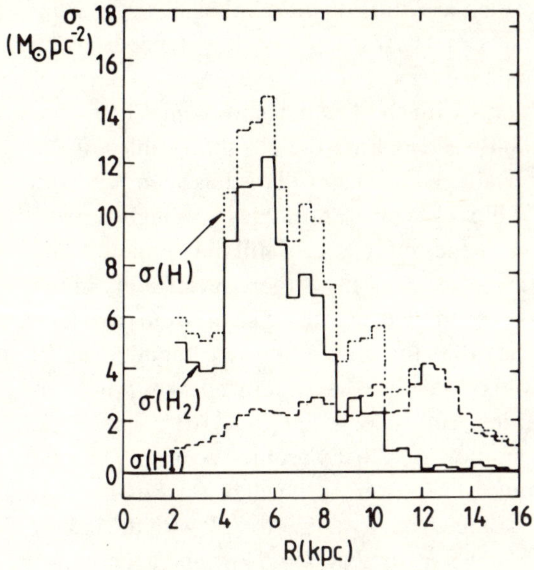

Figure 4.22. Distribution with Galactocentric distance R of the surface densities of atomic and molecular hydrogen, from the work of Gordon and Burton (1976). The sun is at R = 10 kpc. The surface density of molecular hydrogen is necessarily imprecise because of problems with the conversion from the measured CO intensity to molecular hydrogen column density and possible radial dependence of the conversion factor. H: HI + H_2.

1 % by mass of the ISM. Although the distribution of atomic hydrogen (HI) is reasonably well known, from studies of the 21 cm radiation, that of the molecular component (H_2) is the subject of much debate, for the reasons mentioned in Section 4.1.

As an estimate of the large scale distribution of gas, assuming azimuthal symmetry, we can consider the surface densities of HI and H_2 given by Gordon and Burton (1976) and shown in Figure 4.22. By surface density, $\sigma(R)$, we mean $\int_{-\infty}^{\infty} \rho(z)\, dz$ at a fixed Galactocentric radius, R. By way of example, a mean density $\rho = 1$ atom cm^{-3} over a scale height $z = 150$ pc on either side of the plane would give $300 \times 3 \times 10^{18} \approx 10^{21}$ atom cm^{-2}, i.e. $10^{21} \times 1.7 \times 10^{-24} \times (3 \times 10^{18})^2/2 \times 10^{33} \approx 1$ solar mass pc^{-2} ($=1 M_\odot$ pc^{-2}). This plot has been extensively used in γ-ray analyses.

The first realisation of the importance of the then newly discovered H_2 component came from the work of Solomon and Stecker (1974) and Dodds *et al.* (1974). At that time, singularly high surface densities of H_2 had been claimed by Scoville and Solomon (1973) from the CO observations, and the early conclusions are thus of little value now. In what follows, more recent estimates of H_2 will be used but it needs to be stressed that these are still in doubt.

There are two main problems with the CO measurements: first, that they are not yet complete; and second, that the absolute values of the derived H_2 densities are uncertain. The Galactic Plane has been surveyed for CO with large diameter radio telescopes and correspondingly narrow (e.g. 1′) latitude ranges but wide surveys are still incomplete. The problem about the calibration is that not only is there uncertainty in the local conversion for CO → H_2 but there is the strong suggestion (to which the present authors subscribe) that the conversion is a function of R. Adopting a common terminology we write the column density of molecular hydrogen $2N(H_2)$ (in atoms cm^{-2}) as $\alpha \int T(^{12}\text{CO})\,\mathrm{d}V$, where the integral is taken over the temperature–velocity profile. Writing α_{20} as the value of α in units of 10^{20} atom cm^{-2} K^{-1} km^{-1} s, it will be shown later that contemporary local estimates range from $\alpha_{20} = 2.7$ to 7.2, with the lowest (favoured here) being reduced by a factor 2 at $R = 6$ kpc and a factor 3–5 at the GC. In the following, where other work is described, the value of α_{20} adopted will be quoted.

4.6.3 Correlation of gamma-ray intensity with gas column density

Correlations in the outer Galaxy. Many studies have been made of this correlation, both for HI alone and HI + estimated H_2, the studies being made for several different latitude and longitude ranges. The objective is usually to determine the manner in which the intensity of the initiating cosmic rays varies with position in the Galaxy, and thereby to throw light on the cosmic ray origin problem. Specifically, a firm distinction has still to be made between the majority of cosmic ray protons being of Galactic or of extragalactic (EG) origin. The extent to which such a distinction is possible will become clear shortly.

It has long been known that there is only a little molecular gas in the outer Galaxy (second and third quadrants) and, in consequence, this region has been studied rather well. Inspection of Figure 4.6 shows a problem however, namely the γ-ray flux is small and the statistical errors are relatively large.

A number of analyses have been made of the SAS II data, and Figure 4.23 shows the result of an early one.

As in all correlation studies, we write the γ-ray intensity as

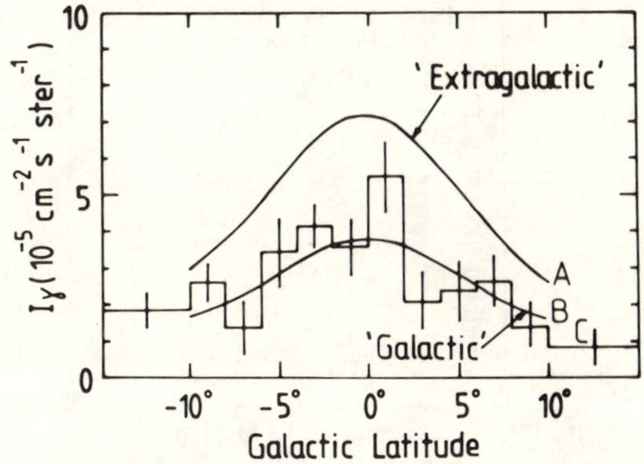

Figure 4.23. Comparison of the latitude distribution of γ-rays above 100 MeV in the Galactic Anti-Centre (SAS II: Fichtel *et al.* (1975)) with the predictions for a Galactic origin of the particles (B) and an extragalactic origin (A). The analysis is from the work of Dodds *et al.* (1975). The data are averages over l: 90°–100°, 160°–170° and 200°–260°; the CRAB direction has been omitted.

$$I_\gamma = \frac{q}{4\pi} N(\mathrm{H}) + I_\mathrm{b} \tag{4.4}$$

where $N(\mathrm{H})$ is the column density of gas in atoms cm^{-2}, q is the emissivity for the energy range in question in γ-rays $atom^{-1}\ s^{-1}$, and I_b is the background flux, a quantity which comprises both instrumental and extragalactic components.

The important quantity of relevance to cosmic ray studies is q, the γ-ray emissivity. Clearly, this quantity is proportional to 'the cosmic ray intensity', or more accurately the appropriate combination of the intensities of its electron and proton components. Under the assumption that the ratio of protons to electrons is independent of position in the Galaxy, an assumption which is only approximately valid, we can revert to using the term 'cosmic ray intensity' and this approximation is followed here. We are, of course, mindful of the fact that the relative contributions of protons and electrons as parents of the detected γ-rays depend on γ-ray energy (see Figures 4.20 and 4.21).

The earliest work relating to q, and the magnitude of the cosmic ray intensity in the outer Galaxy, is that of Dodds *et al.* (1975). Figure 4.23 shows the results of their study using the SAS II data. In the figure, comparison is made of the latitude distribution expected on the basis of a constant cosmic ray intensity at all R (EG origin) and one that follows the

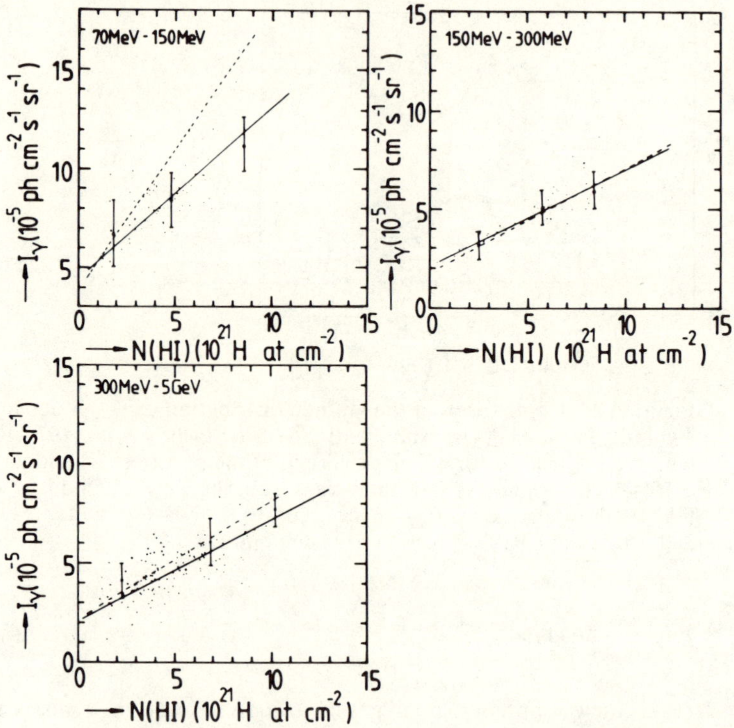

Figure 4.24. Comparison of COS B γ-ray intensities and HI column densities (4° × 4° bins) in the second and third Galactic quadrants. The regions containing point-like sources have been excluded. Typical statistical errors are indicated. Solid lines: least squares fits to the data points; dashed lines: for the local Galactic region as given by Strong *et al.* (1982) using 'high' latitude data ($|b| = 11°–19°$). The near equality of the full and dashed lines for the two upper energy brackets is taken by the authors of this work (Bloemen *et al.* 1984*a*) to indicate that the average cosmic ray gradient in the outer Galaxy is small (or absent) here. See also Figure 4.25.

approximately known radial distribution of SNR (Galactic origin). It is apparent that a Galactic origin is favoured. There is an immediate problem, however, when it is realised (Figure 4.21) that some 30–50 % of the γ-rays in question ($E_\gamma > 100$ MeV) come from electrons. Electrons, certainly being of Galactic origin (see Section 4.6.6), will probably have a steep gradient and it is not inconceivable that the important proton component has little or none. Thus, it is imperative to examine the situation for γ-rays of higher energy, where the cosmic rays responsible are almost entirely protons (and associated heavier nuclei). This will be done shortly.

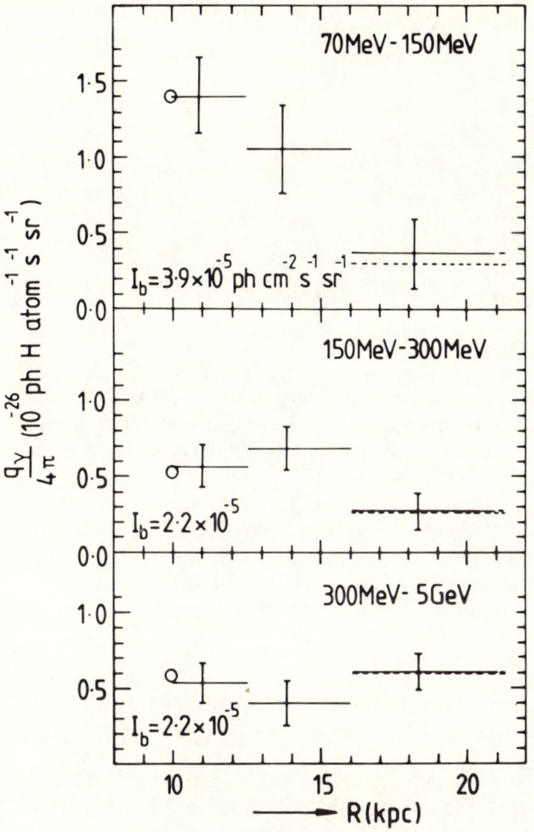

Figure 4.25. Radial distribution of q_γ (i.e. a constant times the cosmic ray intensity) in the outer Galaxy for the three COS B energy ranges, from the work of Bloemen *et al*. (1984*a*). The local emissivity values of Strong *et al*. (1982) are shown for $R = 10$ kpc as circles. Dashed lines: $R > 16$ kpc, showing the q_γ values after correction for a π^0 decay input spectrum.

Returning first to the straightforward correlations of I_γ with N(HI) in the outer Galaxy, the most recent results, which come from the analysis of COS B observations by Bloemen *et al.* (1984*a*), are shown in Figure 4.24. There is seen to be a reassuringly good correlation.

Bloemen *et al*. have gone on to examine the detailed (l, b) dependence of the correlation of I_γ with N(HI) for the situation where N(HI) is divided into radial distance bands (10–12.5 kpc, 12.5–16 kpc and beyond 16 kpc). Regions where there are possible sources (see 2CG catalogue, Table 4.1) are excluded. The resulting q values are shown in Figure 4.25. A rather dramatic result is seen in this figure. Although there is a strong

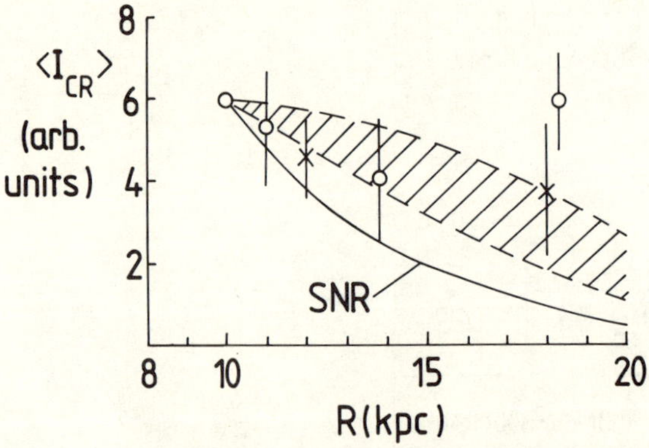

Figure 4.26. Gradient of cosmic ray intensity in the outer Galaxy for COS B data with $E_\gamma > 300$ MeV, namely the regions where protons dominate as the initiating particles. ○: from the analysis of Bloemen *et al.* (1984*a*) (and as shown in Figure 4.25); ×: from an alternative analysis by Bhat *et al.* (1986). The shaded area represents expectation for the situation where the source distribution follows that marked SNR but the flaring of the trapping region causes the particle lifetime to be longer in the outer Galaxy. The limits correspond to extreme assumptions about the diffusion coefficient.

gradient for the lowest energy band (for which the primary particles are almost entirely electrons), that for the second band is smaller, and a gradient in the highest energy band (≈80 % protons) is essentially absent. If true, this is a most important result, suggesting that the protons could well be extragalactic. In view of its importance, Bhat *et al.* (1986) have reanalysed the COS B data for $E_\gamma > 300$ MeV with the results shown in Figure 4.26; it will be seen that there might now be a small gradient. This topic is taken up again in Section 4.6.6 in greater detail.

Correlations at high latitudes. The rather uncertain molecular hydrogen component is confined mainly to the inner Galaxy (Figure 4.22) and, although it is not negligible locally, the uncertainties are not too great. This means that correlation studies of I_γ with gas column densities can be carried out with some confidence. The correlation of I_γ with HI alone has some value, and such early analyses of SAS II data were carried out by Fichtel and colleagues (e.g. Fichtel *et al.* 1978). Figure 4.27 shows the results. It is from these plots, and the corresponding ones for the lower energy bracket, that the extragalactic component of the γ-ray flux has been determined. The correlation is seen to be good.

As remarked already, molecular gas is not completely negligible and a

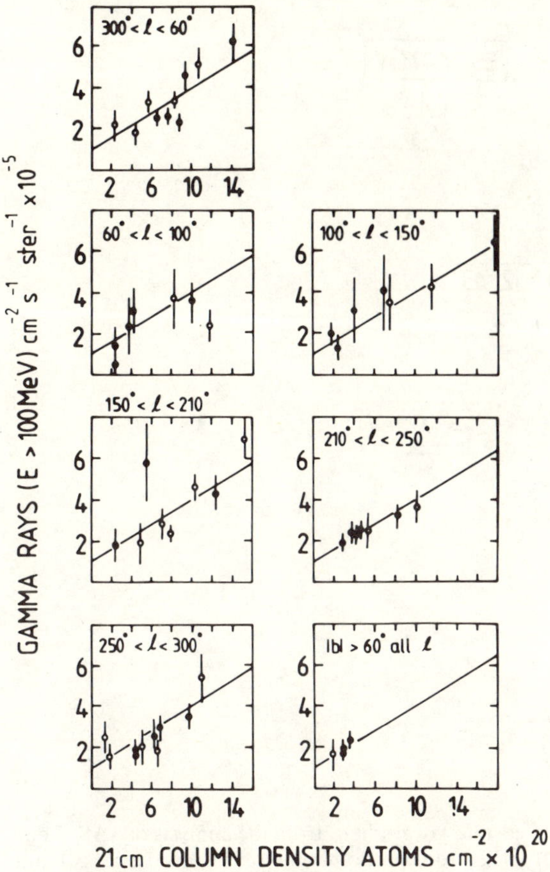

Figure 4.27. SAS II γ-ray intensity vs column density of HI for latitudes $|b|$ greater than 12.8° for the indicated longitude intervals (Fichtel *et al.* 1978). Solid line: best fit to all the data points.

proper allowance for its presence is necessary, particularly in view of the fact that there is some correlation between $N(\mathrm{HI})$ and $N(\mathrm{H_2})$ and this is longitude dependent, i.e. neglect of $N(\mathrm{H_2})$ can lead to incorrect estimates of local cosmic ray intensities. In an attempt to allow for H_2 the galaxy count technique has been used to estimate the total column density of gas ($N(\mathrm{H_t})$) as a function of l and b, i.e. $N(\mathrm{H_t}) = N(\mathrm{HI}) + 2N(\mathrm{H_2})$. The principle of the method is to make the reasonable assumption that the 'density of galaxies' brighter than some limiting magnitude (in number of galaxies per square degree) is constant outside the confines of our own Galaxy. The number per square degree seen at earth will, however, be

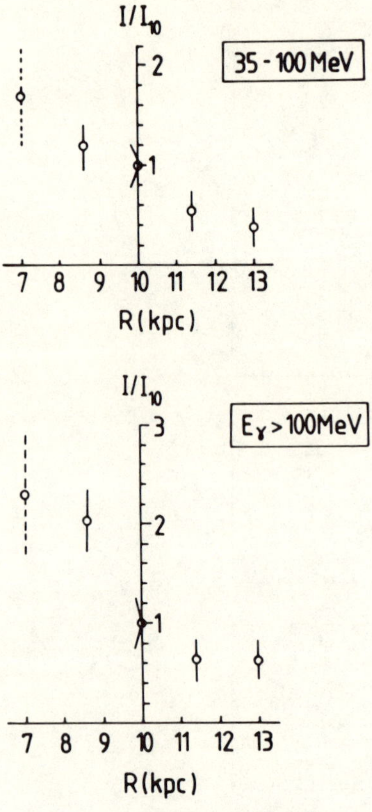

Figure 4.28. Local cosmic ray gradient from the analysis of SAS II data by Issa *et al.* (1981). Electrons predominate for the lower energy band. In the higher band protons probably contribute 50–70 % of the initiating particles. I_{10}: local cosmic ray intensity.

dependent on the column density of dust along the line of sight to the edge of the Galaxy. In turn, if the dust-to-gas ratio is constant, the total column density of gas can be derived. Lebrun and Paul (1983) have used this technique with the SAS II results (as have Issa *et al.* 1981 and Strong and Wolfendale 1981). Lebrun *et al.* (1983) and, more recently, Strong *et al.* (1982), have done the same thing with the COS B observations.

Referring to the range $|b| = 10°–20°$, we can examine the work of Issa *et al.* Here, the relation between galaxy count, N_g (=number of galaxies per square degree in the Lick galaxy catalogue) and the column density of gas as a whole in atoms cm^{-2}, $N(H_t)$, was taken from Strong and Lebrun (1982):

$$N(\mathrm{H_t}) = 2 \times 10^{21} \log\left(\frac{N_g^0}{N_g}\right) \text{ atom cm}^{-2} \tag{4.5}$$

where $N_g^0 = 50$ is the estimated galaxy count extrapolated to the situation of no Galactic absorption.

The result for the smoothed radial distribution is given in Figure 4.28; the apparent gradient for $R > 10$ kpc is in accord with the results shown in Figure 4.23. The rather steep rise in the inner Galaxy is considered later, in Section 4.6.6.

More extensive analysis of the COS B data, undertaken by Strong *et al.*, gives the important results shown in Figure 4.29. There are many interesting features in these plots. There is good general agreement with observation and expectation on the basis of a constant cosmic ray intensity over the region in question, but the gradient claimed in Figure 4.23 shows through as an excess of observation over expectation in the inner Galaxy and a deficit in the outer. The nearby molecular clouds in Orion and Taurus are clearly visible. There is an interesting reduction in the predicted γ-ray intensity in the direction of the Perseus cluster of galaxies; this is quite certainly due to the fact that the value of N_g in this direction is excessive (because of the cluster) and the result is an underestimate of $N(\mathrm{H_t})$. Inevitably this sensitivity to galaxy clustering is a shortcoming of the galaxy count method.

Continuing in a critical vein, the method is also sensitive to the undoubted variations from place to place of the dust-to-gas ratio (it is the dust which absorbs the light and thus reduces the number of galaxies seen along a particular line of sight). Strong *et al.* (1982) use the differences in observed and expected γ-ray flux for those regions where galaxy counts are not available, $220° < l < 10°$, to estimate the column densities of molecular gas, but this seems to be a rather risky procedure.

Table 4.4 lists the q values derived for the high latitude regions; it will be seen that in some cases H_2 has been disregarded and in others it is included in $\mathrm{H_t}$.

Correlation with gas in the inner Galaxy. This is the region where the biggest uncertainty arises because of the problems already mentioned concerning the lack of data on molecular gas in some regions and the general difficulty of converting such information as is available (from CO) into the column densities of molecular hydrogen.

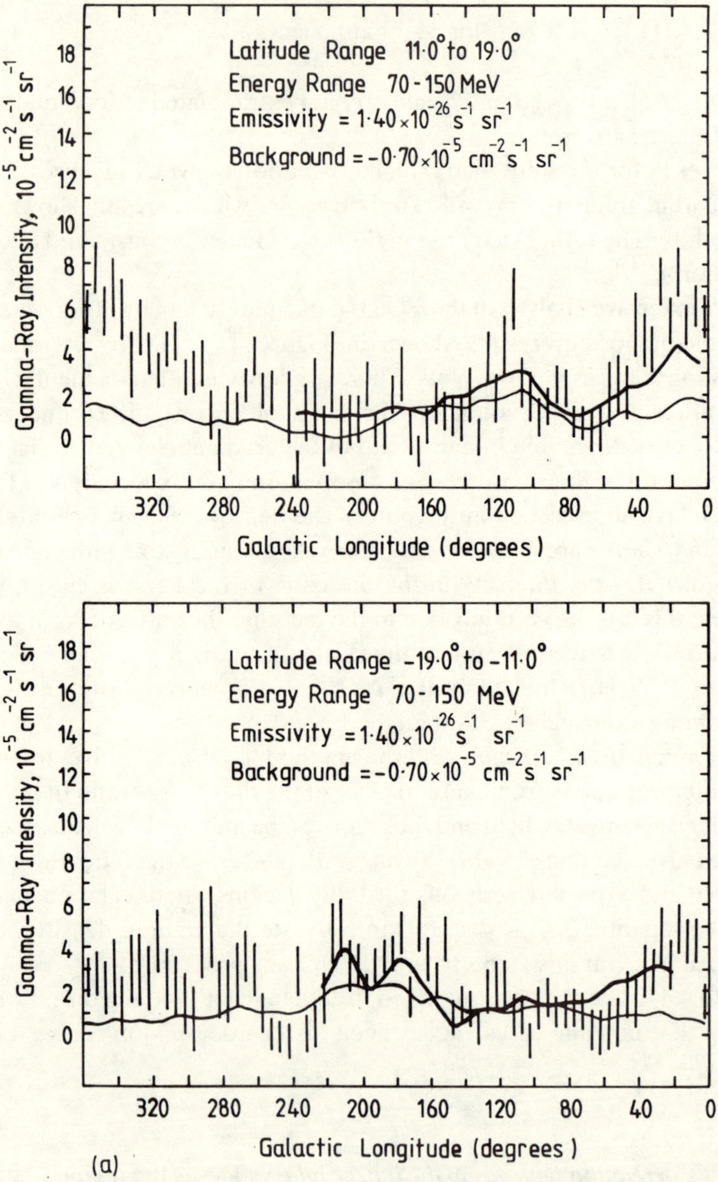

Figure 4.29. Comparison of measured and predicted γ-ray intensities for COS B (Strong *et al.* 1982). Lower line: corresponds to the contribution from interactions with atomic hydrogen alone. Upper line: relates to all gas, the column densities being derived from the galaxy-count method.

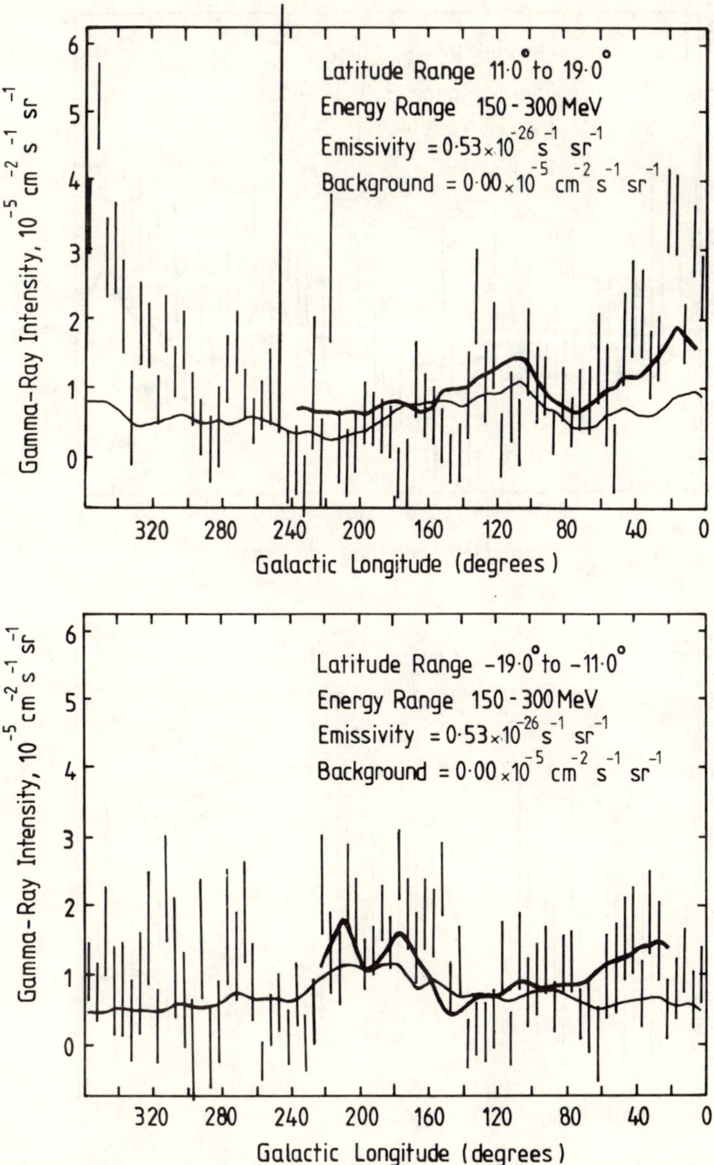
Latitude Range 11·0° to 19·0°
Energy Range 150 - 300 MeV
Emissivity = 0·53 × 10⁻²⁶ s⁻¹ sr⁻¹
Background = 0·00 × 10⁻⁵ cm⁻² s⁻¹ sr⁻¹
Gamma-Ray Intensity, 10⁻⁵ cm⁻² s⁻¹ sr⁻¹
Galactic Longitude (degrees)
Latitude Range −19·0° to −11·0°
Energy Range 150 - 300 MeV
Emissivity = 0·53 × 10⁻²⁶ s⁻¹ sr⁻¹
Background = 0·00 × 10⁻⁵ cm⁻² s⁻¹ sr⁻¹
Gamma-Ray Intensity, 10⁻⁵ cm⁻² s⁻¹ sr⁻¹
Galactic Longitude (degrees)

(b)

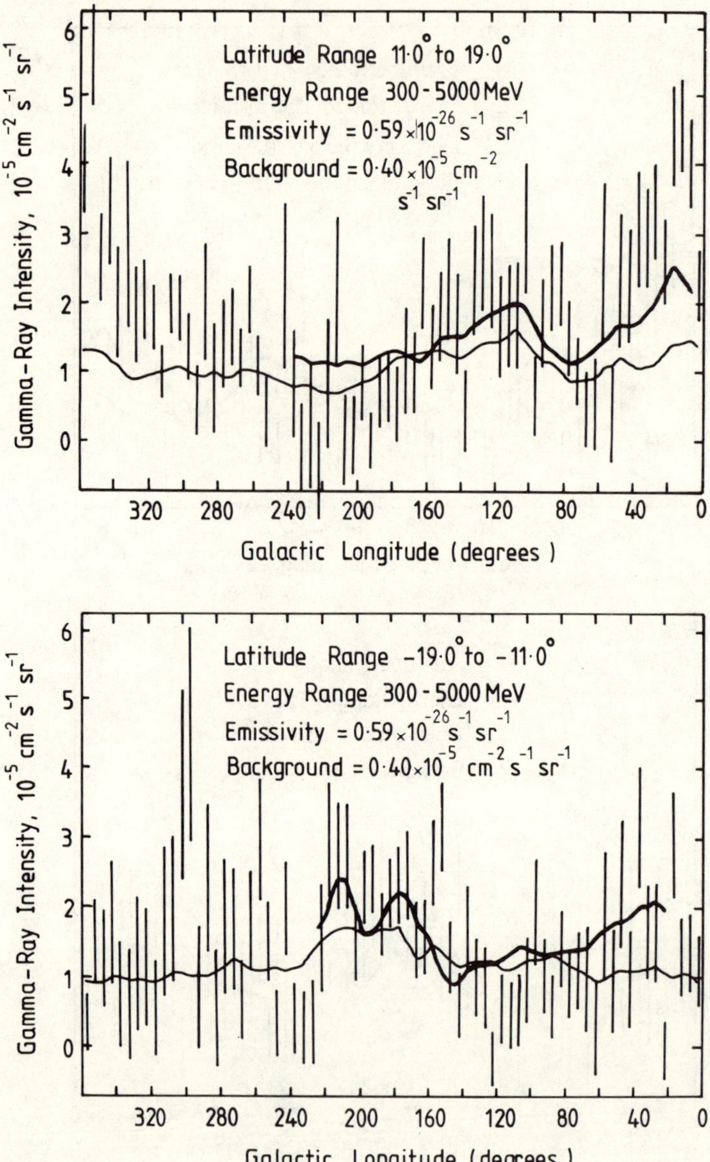
Latitude Range 11·0° to 19·0°
Energy Range 300 - 5000 MeV
Emissivity = 0·59×10^-26 s^-1 sr^-1
Background = 0·40×10^-5 cm^-2 s^-1 sr^-1
Gamma-Ray Intensity, 10^-5 cm^-2 s^-1 sr^-1
Galactic Longitude (degrees)
Latitude Range −19·0° to −11·0°
Energy Range 300 - 5000 MeV
Emissivity = 0·59×10^-26 s^-1 sr^-1
Background = 0·40×10^-5 cm^-2 s^-1 sr^-1
Gamma-Ray Intensity, 10^-5 cm^-2 s^-1 sr^-1
Galactic Longitude (degrees)

(c)

Table 4.4. *Summary of q-values from the work of Strong and Wolfendale (1981)*

Values found from correlations of I_γ with $N(\mathrm{HI})$ are generally higher than for I_γ versus $N(\mathrm{H_t})$ because of the effect of including H_2 in the latter.

Reference	Experiment	l (degrees)	b (degrees)	Linear fit to	$(q/4\pi)/(10^{-26}\ \mathrm{s^{-1}\ sr^{-1}\ atom^{-1}})$ 35–100 MeV		>100 MeV
Fichtel *et al*. (1978*a*)	SAS-2	0–360	$\lvert b\rvert > 12.8$	$N(\mathrm{HI})$	4.3 ± 1.5		3.0 ± 0.8
Lebrun and Paul (1979)	SAS-2	0–360	$\lvert b\rvert > 10$	$N(\mathrm{HI})$	4.5 ± 0.3		3.5 ± 0.2
			$\delta > -25$	$N(\mathrm{H_t})$	3.0 ± 0.2		2.0 ± 0.15
Strong and Wolfendale (1981)	SAS-2	0–240	$20 > \lvert b\rvert > 10$	$N(\mathrm{HI})$	$q_1 = 3.2 \pm 1.0$		$q_1 = 3.1 \pm 0.5$
				$N(\mathrm{H_2})$	$q_2 = 4.0 \pm 1.5$		$q_2 = 1.9 \pm 0.6$
			$\delta > -25$	$N(\mathrm{H_t})$	3.7 ± 1.0		2.6 ± 0.5
Assumed sensitive area (this table)*					$40\ \mathrm{cm^2}$		$59\ \mathrm{cm^2}$
					70–150 MeV	150–300 MeV	300–5000 MeV
Lebrun *et al*. (1983)	COS-B	10–240	$20 > \lvert b\rvert > 10$	$N(\mathrm{HI})$	1.15 ± 0.17	0.76 ± 0.10	0.84 ± 0.08
			$\delta > -25$	$N(\mathrm{H_t})$	1.45 ± 0.15	0.54 ± 0.07	0.46 ± 0.07
Assumed sensitive area (this table)					$25.1\ \mathrm{cm^2}$	$41.1\ \mathrm{cm^2}$	$47.0\ \mathrm{cm^2}$

Assumed conversion for galaxy counts: $N(\mathrm{H_t}) = 2.0 \times 10^{21}\ \lg(50/N_g)\ \mathrm{cm^{-2}}$.

* Except for Fichtel *et al*. (1978).

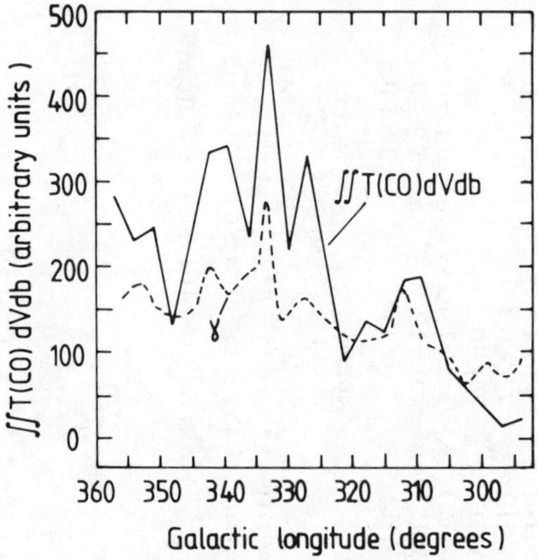

Figure 4.30. Comparison of the approximate column density of molecular hydrogen (assuming $N(H_2) = \alpha \int T(\mathrm{CO})\,\mathrm{d}V$) over $\pm 1°$ in latitude and the γ-ray flux above 300 MeV for $|b| < 1°$ from the COS B data of Mayer-Hasselwander *et al.* (1982). The comparison is from the work of Riley *et al.* (1984).

At the superficial level, there exists general correlation of γ-ray flux with the integral of CO temperature, over the relevant Galactic velocity range, i.e. $\int T\,\mathrm{d}V$. An example is shown in Figure 4.30, where the $\int\int T(\mathrm{CO})\,\mathrm{d}V\,\mathrm{d}b$ values are from a preliminary survey of CO in the southern hemisphere and the γ-ray data are nominally for $|b| < 1°$ and come from the COS B results of Mayer-Hasselwander *et al.* (1982). The word 'nominally' is used because even for energies above 300 MeV the PSF is such that a significant fraction of the γ-rays with latitude actually within $\approx 1°$ are recorded as having $|b| > 1°$. Nevertheless, the comparison is useful and it can be seen that the peaks of $\int\int T\,\mathrm{d}V\,\mathrm{d}b$ and I_γ are in similar positions.

An important fact that follows from these comparisons is that the initiating cosmic rays appear to see the bulk of the molecular gas, i.e. that the particles are not excluded from the gas clouds. The coincidence of the peaks in both Figures 4.30 and 4.31, particularly for the Cygnus region in the latter ($-5.5° < b < 5.5°$, $l \approx 80°$), where the gas is comparatively close to the solar system, supports this contention.

There can be no question at this stage of the near agreement of observation and expectation in Figure 4.31 being used to demonstrate that the cosmic ray intensity is constant in the inner Galaxy. It will be

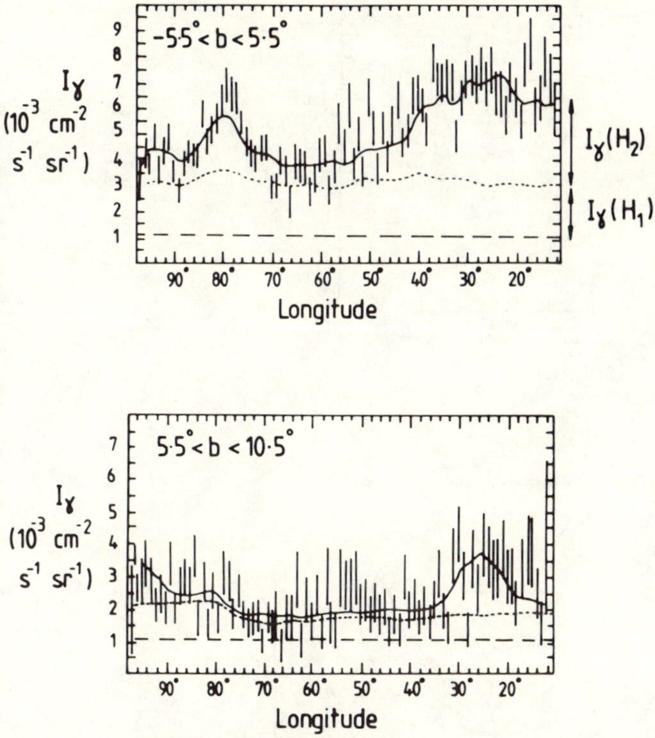

Figure 4.31. Longitude profiles of the observed γ-ray emission for $E > 300$ MeV compared with that expected from H_2 + HI (upper line) and HI alone (lower line), after Lebrun *et al.* (1982). Dotted line: background level.

apparent, from the remarks made in the earlier part of this section, that at least for γ-ray energies below 300 MeV there is evidence that the initiating particles have an intensity gradient in the outer parts of the Galaxy, and on any reasonable model of the cosmic ray sources a similar trend (an increase) would be expected in the inner Galaxy. This topic is taken up again in Section 4.6.6.

4.6.4 Gamma rays from specific molecular clouds

Eventually, when γ-ray data of great precision become available, the study of γ-ray intensity profiles from known molecular clouds will be of great value in examining in detail the interiors of those clouds which are inert, as well as in studying cosmic ray production in the clouds which have sources within them. Although such information is not yet available at least a start has been made.

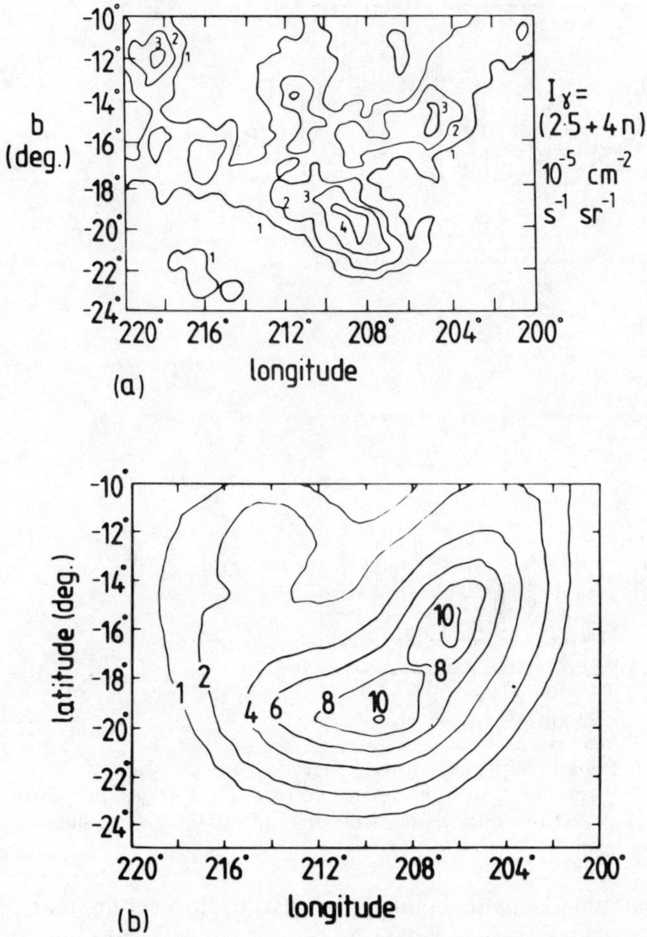

Figure 4.32. The Orion molecular cloud region. (*a*) γ-rays above 100 MeV from the COS B collaboration (private communication, 1984). (*b*) CO contours smoothed to the resolution of the COS B data; the basic CO data are from Kutner *et al*. (1981) and have been smoothed by Houston and Wolfendale (1985). *n* is the contour number.

This topic has already been touched on in Section 4.5.3 in connection with the possibility of distant cosmic ray irradiated clouds simulating the so-called γ-ray sources. Here, we deal with the subject in more detail and start with the nearby clouds in the Orion region, clouds which are so close that their structure is revealed even with the poor angular resolution of the contemporary γ-ray detectors.

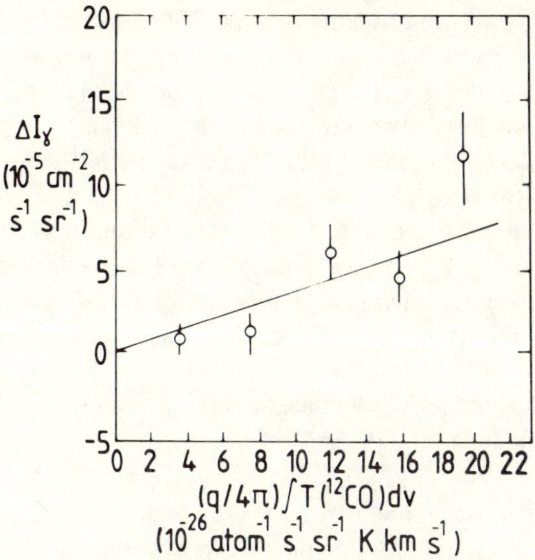

Figure 4.33. Correlation of ΔI_γ (= I_γ observed) − $(q/4\pi)N$(HI) − I_γ (background), with $(q/4\pi) \times (\int T(^{12}\text{CO})\,dV)$ for the Orion molecular clouds (see Figure 4.32). $(q/4\pi)$ is the emissivity per hydrogen atom and N(HI) is the column density of atomic hydrogen. The reasonable correlation indicates that cosmic ray particles are not excluded from regions of the clouds where a significant fraction of the mass resides.

The SAS II data for the Orion region were analysed by Wolfendale (1980) and it was shown that the γ-ray flux could be understood in terms of cosmic ray irradiation of these clouds. The later COS B data, with its improved statistical precision (Caraveo *et al.* 1980, 1981), has allowed the correlation of γ-ray intensity with gas density to be examined in greater detail.

Figure 4.32 shows the contours of γ-ray intensity, from COS B for the Orion region, together with the corresponding smoothed ^{12}CO velocity-integrated antenna temperature ($\int T(^{12}\text{CO})\,dV$), i.e. the quantity proportional to the column density of molecular hydrogen. Figure 4.33 shows the correlation of the component of γ-ray intensity attributable to cosmic ray interactions with the H_2 gas (i.e. total minus background and minus the product of HI interactions) as a function of $\int T(^{12}\text{CO})\,dV$. The near linearity is an important feature in that it shows that the initiating cosmic rays penetrate the bulk of the gas in the clouds and indicates that γ-rays can be used as a probe of gas in various regions of the Galaxy.

Houston and Wolfendale (1985) and Bloemen *et al.* (1984*b*) have both examined the correlation for Orion and used it to derive the relationship between the H_2 column density and $\int T(^{12}CO)\,dV$, making the reasonable assumption that the cosmic ray intensity is the same in the Orion clouds (about 500 pc away) as that locally. Clearly, if the cosmic ray intensity is uniform across the clouds (as the Houston and Wolfendale analysis suggests) this is a meaningful approach. Due to different assumptions, mainly about the γ-ray emissivity, the two groups derive somewhat different results for the conversion factor, α_{20}. Writing $\alpha_{20} = N(H_2)/\int T(^{12}CO)\,dV$ in units of 10^{20} atom $cm^{-2}\,K^{-1}\,km^{-1}\,s$, the values are:

$\alpha_{20} = 3.7 \pm 1.0$ Houston and Wolfendale (1985)
$\alpha_{20} = 5.2 \pm 2.4$ Bloemen *et al.* (1984*b*)

Other clouds have also been examined by the γ-ray technique. As mentioned earlier (Section 4.5.6), the cloud complex in Rho Ophiucus has been extensively studied. Knowledge about molecular clouds from the other branches of astronomy is still rudimentary (although probably better than the γ-ray information, at least in terms of statistical precision), but an analysis was attempted by Issa and Wolfendale (1981*b*) for a number of clouds. A variety of ^{12}CO data were used (much of it from Blitz 1980) together with the SAS II catalogue of Fichtel *et al.* (1978) and initial COS B results. An attempt has also been made now (Figure 4.34) to update the analysis by adopting the more recent COS B observations.

A prominent feature of the results is the presence of a number of clouds for which the observed γ-ray fluxes appear to need a considerable enhancement of cosmic ray intensity in these clouds. This feature has been noted by a number of workers (particularly for Eta Carinae, e.g. Montmerle 1979, 1982), although it is necessary to sound a note of caution that the identification of the γ-ray peaks with distant giant molecular clouds is fraught with difficulty. These 'coincidences' are approximate, spatially, and unlike in the case of Orion the distant clouds appear 'point-like' within the poor angular resolution of the present instruments. Furthermore, it is not inconceivable that there are discrete sources of γ-rays (pulsars?) embedded within the clouds, as distinct from sources of cosmic ray particles.

The importance of the high cosmic ray enhancement factors (F values) seen in Figure 4.34, to the question of the origin of cosmic rays, is taken up in Section 4.6.6.

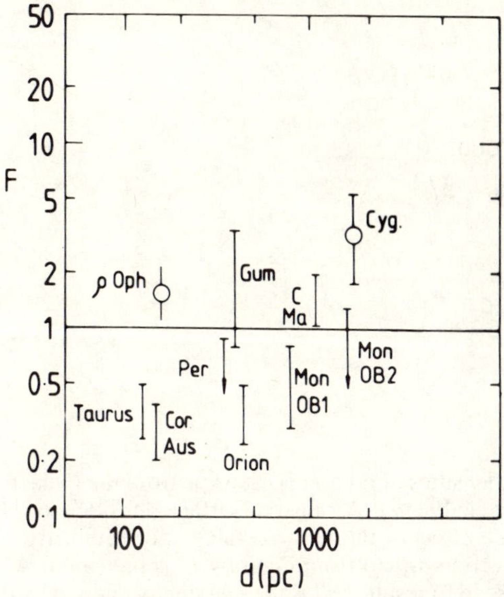

Figure 4.34. Enhancement of cosmic rays in comparison with the local value (F values) for a number of molecular clouds. The plot is an update of work by Issa and Wolfendale (1981*a*, *b*). O: clouds which have been (tentatively) identified with γ-ray sources in the COS B 2CG catalogue of Hermsen (1981). (Despite the update there are still problems with the cloud masses.) d is distance from the sun.

4.6.5 Relevance of cosmic gamma-rays to the mass of gas in the Galaxy

The interrelation of cosmic rays, gas and γ-rays has been stressed already and recent work has enabled interesting conclusions to be drawn about the mass of gas in the Galaxy.

It might be thought that the embryonic subject of γ-ray astronomy, with its statistically small sample of γ-rays ($\approx 10^5$ only), of poor energy and angular resolution, would not be able to compete with the older subject of radio astronomy and the data of high precision obtainable there. There is a problem, however, with radio estimates of the mass of gas in molecular clouds, and this is associated with the difficulty in relating the observations of CO to densities of H_2. Interstellar chemistry is not only a difficult subject in its own right, with vast numbers of possible reaction to contend with, but the physical conditions inside the clouds (temperature, degree of ionisation, etc.) are poorly known. As a consequence, in the authors' view, the γ-ray results on the gas content of the Galaxy is probably as good as, and possibly better than, that available from radio.

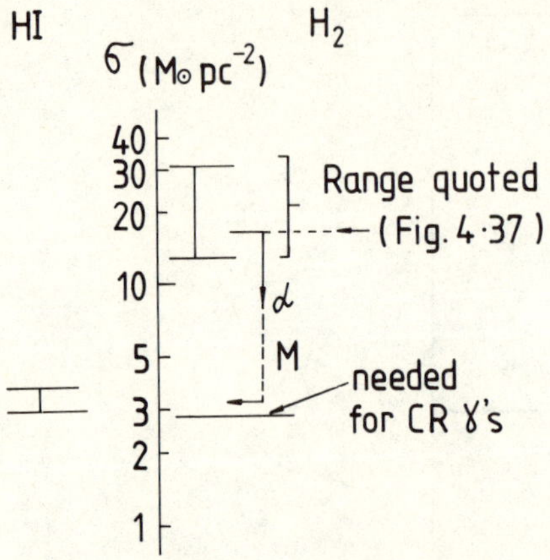

Figure 4.35. Surface densities of HI and H_2 at 6 kpc from the Galactic Centre (after Bhat *et al.* 1984*b*). A range of surface densities for H_2 have been quoted; the effect of the new α_{20} value and metallicity correction ($M'(6) = 2$) on $\sigma_{H2}(6)$ from Figure 4.37 is indicated. A consistent picture is seen to result. (M' is the metallicity divided by the local value.)

Bhat *et al.* (1984*b*) have examined the problem recently and their method will be followed here. The argument is in two parts. Examination of the local giant molecular clouds, particularly those in Orion, yields a value of α_{20} which is a factor of about two smaller than that used by some radio astronomy groups – e.g. Sanders *et al.* (1984) – to be referred to as SSS. Arguments are then put forward for a continuation towards the inner Galaxy, of the low energy cosmic ray gradient determined in the outer Galaxy (see Figure 4.25 and Section 4.6.3) and this appears to require an even lower value of α_{20} in the inner Galaxy. Arguments based on interstellar chemistry, such as it is understood, suggest that $\alpha_{20}(R) \propto (M(R))^{-1}$, where M is the 'metallicity', namely the ratio of density of atoms such as oxygen to that of hydrogen. It is well known (e.g. Pagel and Edmunds 1981) that there is a metallicity gradient in the Galaxy with M(6 kpc) about twice that at $R = 10$ kpc. The same argument, but with less detail, was advanced earlier by Blitz and Shu (1980).

Figure 4.35 shows the situation at $R = 6$ kpc. It is found that adopting $\alpha_{20}(R) \propto (M(R))^{-1}$ yields a γ-ray emissivity there roughly as observed. A measure of confirmation comes from an analysis of γ-rays from the GC – a topic of interest to the present work in its own right. The metallicity here is $M(R = 0) \approx 3$–5 times the local value and the ensuing inferred gas mass appears to give a much more reasonable result for the cosmic ray intensity in this region. Figure 4.36 summarises the situation.

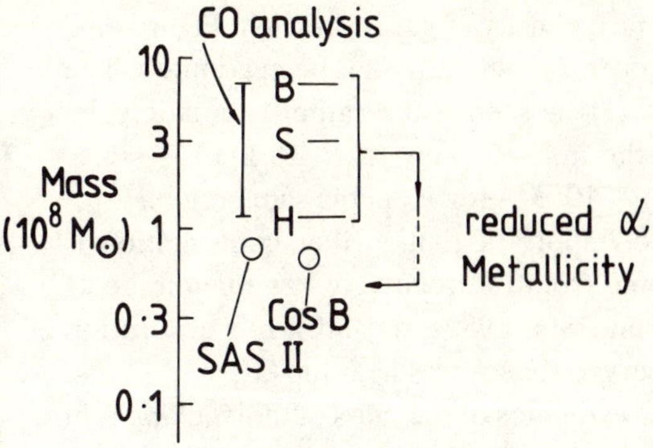

Figure 4.36. Mass of gas within 400 pc of the Galactic Centre, after Bhat *et al.* (1984*b*, 1985*a*). B, S, H: estimates by Bania (1977), Scoville *et al.* (1974), and Heiligman (1982), respectively, using different α-values. ○: what would be expected from the γ-ray observations if the cosmic ray intensity were the same at the Galactic Centre as locally. The B, S, H masses would require unreasonably low cosmic ray intensities at the GC. Adoption of the reduced α-value and $M'(0) = 3$ results in a mass ($\approx 5 \times 10^7$ $M_\odot$), which would correspond to a cosmic ray intensity at the GC a factor ≈ 2 times that locally – a much more reasonable result.

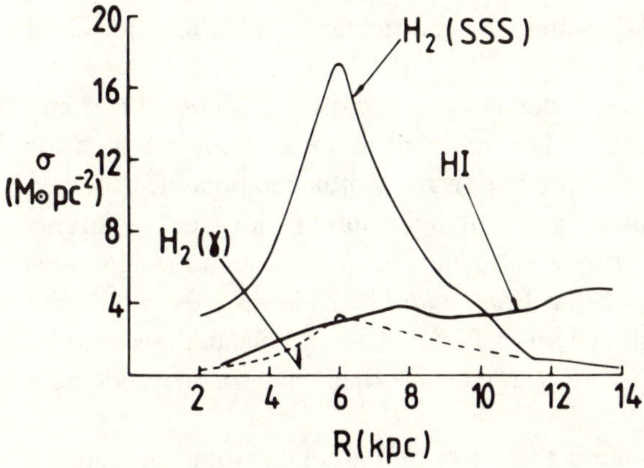

Figure 4.37. Dependence of the surface density of molecular and atomic hydrogen as a function of Galactocentric radius under the assumption of Galactocentric symmetry. SSS: from the CO analysis by Sanders *et al.* (1984). HI: mean of estimates by Gordon and Burton (1976) and Li *et al.* (1983). $H_2(\gamma)$: from the γ-ray estimate of Bhat *et al.* (1984*b*, 1985*a*).

Finally, the effect on the surface density of gas is shown in Figure 4.37 where, as is customary, Galactocentric symmetry has been adopted. The reduction in surface density of H_2 is seen to be rather dramatic. The corresponding mass of H_2 in the inner Galaxy ($R < 10$ kpc) is $\approx 6 \times 10^8\ M_\odot$, to be compared with $1 \times 10^9\ M_\odot$ for HI in the same region.

The topic of gas mass is so important that further confirmation is required and this seems to come from the relatively low magnitude of both X-ray absorption (Blitz and Shu 1980) and infra-red absorption (Eaton *et al.* 1984) that are observed for sources near the GC.

The implications of the lower estimates of the mass of molecular gas in the Galaxy – for subjects as diverse as star formation processes and the stability of the Oort cloud of comets – are the subject of contemporary work.

4.6.6 Relevance of cosmic gamma-rays to the origin of the cosmic radiation

Unlike cosmic ray particles, cosmic γ-rays travel in straight lines and, insofar as much of the γ-ray flux is caused by cosmic ray particles interacting with gas in the ISM, one would hope for some clues to be given with regard to the question of cosmic ray origin.

These studies are relevant in two main ways: (i) the information provided by large scale intensity gradients, and (ii) the cosmic ray enhancement factors (F values) for particular objects and localised regions.

The large scale intensity gradients were discussed in Section 4.6.3, and it was pointed out that there is agreement about a strong gradient for cosmic ray electrons. As for the important proton component, Bloemen *et al.* (1984*a*) find no intensity gradient in the outer Galaxy for the parents of the highest energy γ-rays (300–5000 MeV) – see Figure 4.25. This conclusion is so important that Bhat *et al.* (1986) have re-examined the situation, with the result shown in Figure 4.26. If this analysis is to be believed, the situation is now equivocal – there may or may not be a modest gradient.

A natural claim to make if there were no gradient would be that the protons are mainly extragalactic, but we believe that this is not yet warranted. The reason is that particle diffusion can reduce the gradient considerably. In the absence of a knowledge of the magnitude of the diffusion coefficient $D(R, z, E)$ and its dependence on radial distance, R, scale height, z, and energy, E, and the nature of the Galactic 'container', accurate calculations cannot be made of the relationship between the distribution of sources and the ensuing distribution of the CR intensity. However, Bhat *et al.* (1986) have made an order of magnitude estimate by assuming simple diffusion in the z-direction to a boundary whose height is proportional to the scale height of HI in the outer Galaxy. Limiting forms are taken for $D(R)$, and the result is shown by the hatched area in Figure 4.26. It is seen that the γ-ray derived cosmic ray gradient is not inconsistent with expectation, although at this stage the question whether a gradient is there or not must still remain open.

Work has also been done on the γ-ray data for the inner Galaxy and mention can be made here of the analysis by Kniffen *et al.* (1983). These authors made the assumption that the cosmic ray intensity, I_{CR}, is proportional to the density of total gas and chose a value of α_{20}(CO $\rightarrow$ H_2 conversion) to give consistency for the total intensity. Understandably, a gradient of cosmic ray intensity resulted, since the initial assumption of $I_{CR} \propto \rho_{gas}$ implies a Galactic origin, but the assumption of α_{20} being R-independent is *ad hoc*. In a similar way, the earlier analysis of Issa *et al.* (1981) in which, for comparatively local regions, correlation of I_γ with N(HI) was studied (the neglect of H_2 is not catastrophic because the correlation of H_2 with HI is not strong) gave results which are useful but not compelling.

Better information comes from the detection of local gradients (Bhat *et al.* 1984*a*) and from the recent suggestion that the very considerable local gradient (Figure 4.38) may be due to an enhanced cosmic ray intensity in the Loop I local SNR. Although this suggestion has yet to be confirmed, it can be considered here, if only as an example of the manner in which γ-ray astronomy will eventually contribute to the problem of probing the important SNRs from the cosmic ray point of view.

It is likely that the North Galactic Spur (=Loop I radio feature) should be identified with the remnant of an SN which is centred on a point some 130 pc away from the sun and has at present a radius of 115 pc (Berkhuijsen *et al.* 1971). Although it was originally thought that the initial energy of the SNR needed to be much greater than the value of 10^{51} erg typical of conventional SN, recent SN models (e.g. by Blandford and Cowie 1982) can explain many features with only a typical initial explosion. Blandford and Cowie predict a γ-ray intensity from the SNR surface of order $7 \times 10^{-6}\,\text{cm}^{-2}\,\text{s}^{-1}\,\text{sr}^{-1}$ and, remarkably, this is close to the estimate made for the Loop I remnant by Bhat *et al.* (1985*b*) from their analysis of the SAS II and COS B data. Figure 4.39 summarises the situation.

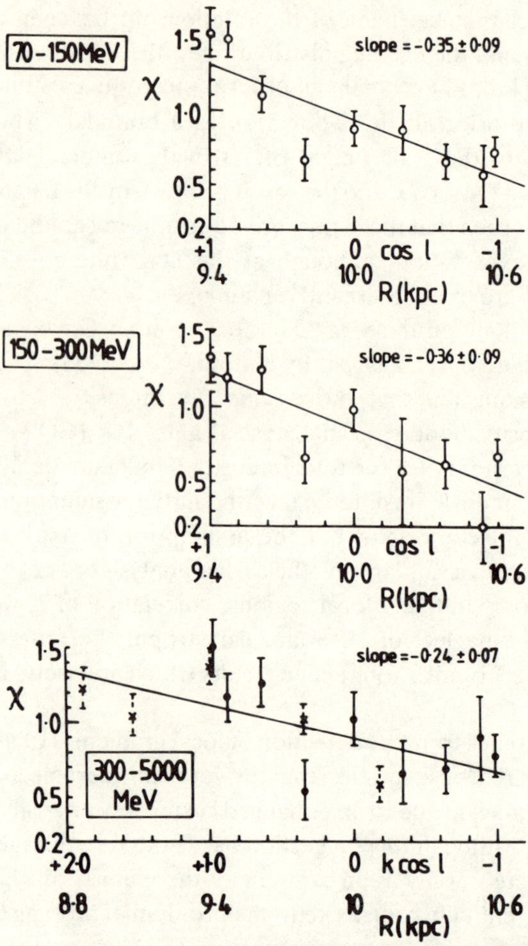

Figure 4.38. Local cosmic ray gradient derived from an analysis of COS B data by Bhat *et al.* (1984*a*). The latitude range is $|b| = 11°–19°$ (and for $b = 5.5°–10.5°$ in the highest energy region – shown dotted). χ: ratio of observed γ-ray intensity to that expected for a constant cosmic ray intensity. l: longitude; R: Galactocentric radius; $k = 1$ for $|b| = 11°–19°$ and 2 for $5.5°–10.5°$. The most important feature is the slope for the highest energy band, where the initiating particles are mainly protons. This strongly suggests that the protons are produced in Galactic sources rather than being incident on the Galaxy from outside. Bhat *et al.* (1985*b*) make the case for some of this gradient being due to an enhanced cosmic ray intensity in the Loop I SNR (see Figure 4.39).

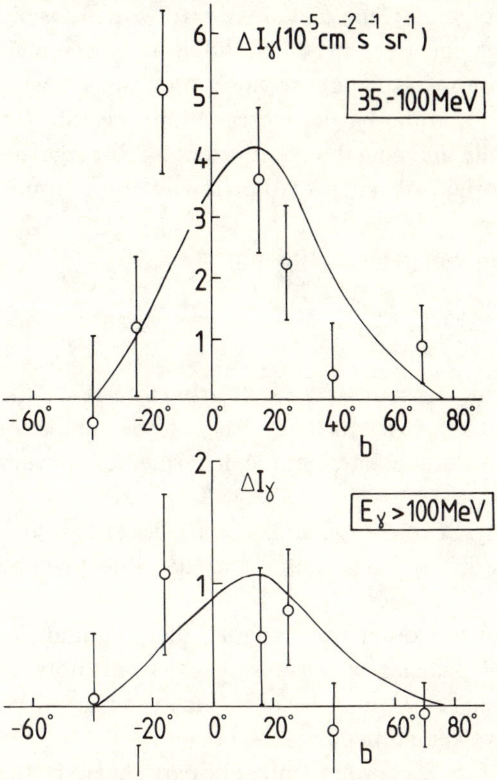

Figure 4.39. Excess γ-ray intensity associated with the Loop I SNR, from Bhat *et al.* (1985*b*). SAS II data on γ-ray intensities were used. Cuts are taken across the remnant and ΔI_γ is the difference between the excess of observation over expectation for a cosmic ray intensity within the remnant to that outside the remnant (this technique minimises the effect of systematic errors on the expected values). Full line: approximate expectation for the case where the cosmic ray intensity is uniform within the remnant (in fact it clearly is not, but the effect on the shape is small). Electrons predominate as parents in the 35–100 MeV region and protons for $E > 100$ MeV. COS B data are only available at present for $|b| < 19°$, but here they give strong support to the contention that there is an excess at all γ-ray energies.

An examination of the energetics yields a figure of $\approx 10^{50}$ erg in cosmic rays inside the remnant at the present time and if it is assumed that remnants of this dimension ($\approx$100 pc) are close to the stage where cosmic ray confinement fails and the particles escape into the ISM (or perhaps – and this is essentially the same thing – adjacent SNR merge) an acceptable theory for the origin of much of the (low energy) cosmic radiation results.

This is clearly one of the growth points of the subject.

4.7 Extragalactic gamma-rays

4.7.1 General remarks

The poor statistical precision of γ-ray results from SAS II and the background problems with COS B mean that, although the subject of extragalactic γ-rays is of considerable interest, the uncertainties are very great.

As with the Galactic γ-rays, two components can be discerned: 'discrete' sources and a general diffuse background, and the subject will be dealt with in that order.

By analogy with the results at other wavelengths, the potential EG sources comprise individual galaxies of various types, from 'normal' galaxies to quasars. Galaxy clusters should also be included and, as will be demonstrated, examples have been found in each class.

Again, by analogy, the diffuse background might be expected to be the subject of great debate as concerns the fraction of the flux to be identified with unresolved discrete sources as distinct from a *bona fide* diffuse component; it is.

4.7.2 Discrete sources of extragalactic gamma-rays

At the energies in question, the number of identified EG objects is small at present.

Most important is the almost certain observation of the 'nearby' quasar 3C273. Figure 4.40 summarises the measurements above 1 keV. Inspection of the 2CG catalogue (Table 4.1) shows that this object appears therein and this is the only source of EG origin definitely detected in the COS B analysis. Upper limits for many other potential sources have been given by Pollock *et al.* (1981); the conclusion that is drawn about the form of the spectrum of EG sources (active galaxies) is that it must steepen significantly between the X-ray and high energy γ-ray regions (Bignami *et al.* 1979).

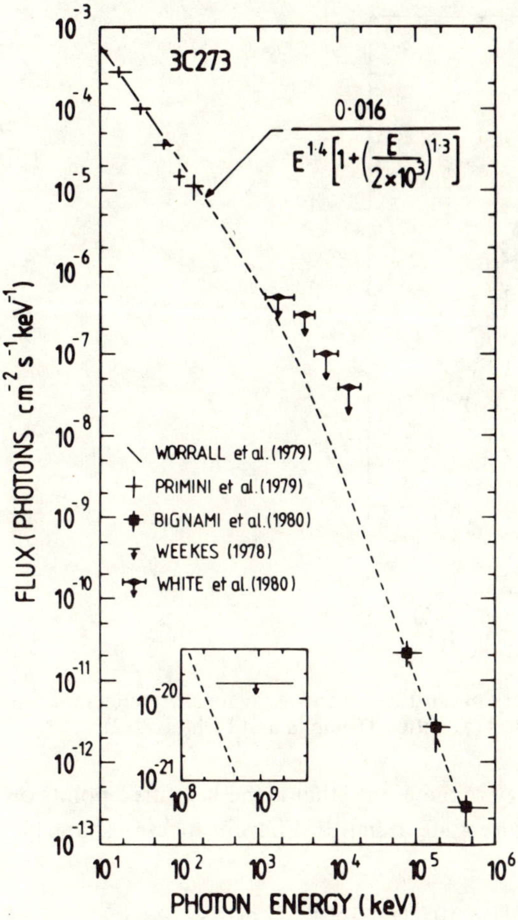

Figure 4.40. Summary of hard X-ray and γ-ray measurements for the quasar 3C273 (after Trombka and Fichtel 1982). The sources of the intensities are given in the inset.

Returning to 3C273, inspection of the smooth curve which gives a reasonable fit to the experimental intensities in Figure 4.40 shows that the peak luminosity (in terms of energy content per decade of photon energy) is in the MeV region. This is a remarkable result and is an indication of the initial energy production process being efficient at rather high energies. The attendant increase in spectral slope above about 2 MeV is interesting, and its possible significance will be discussed later.

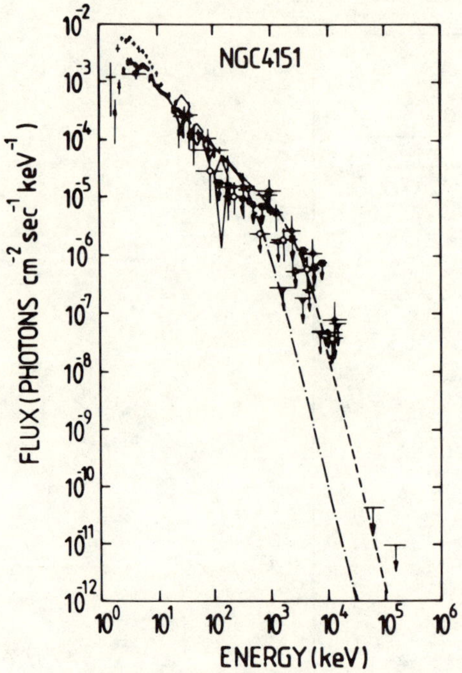

Figure 4.41. Summary of hard X-ray and γ-ray measurements for the Seyfert Galaxy NGC 4151 (after Trombka and Fichtel 1982).

Taking the COS B energy range alone, that is the last three points on the energy spectrum in Figure 4.40, a slightly different spectral shape has been given:

$$N(E_\gamma) = (3.7 \pm 1.4) \times 10^{-6}(E_\gamma/150)^{-2.5}\ \mathrm{cm^{-2}\,s^{-1}\,GeV^{-1}} \qquad (4.6)$$

where E_γ is the energy in MeV (Bignami *et al.* 1981); but the difference from the form given in Figure 4.40 is not significant.

At X-ray energies there is a well known variability of output but none has been detected so far in the γ-ray energy region, possibly because the γ-ray data are so sparse.

The nearest quasar is QSO 0241 + 622 and this is tantalisingly close to the COS B source 2CG 135 + 01. However, there is also a Galactic source nearby (GT 0236 + 61) and identification is not clear cut (Dean and Ramsden 1981).

Coming down a little in energy scale we reach the Seyfert galaxies. These galaxies are prominent emitters at X-ray and longer wavelengths and might be expected to show up in γ-rays too. Emission has been found

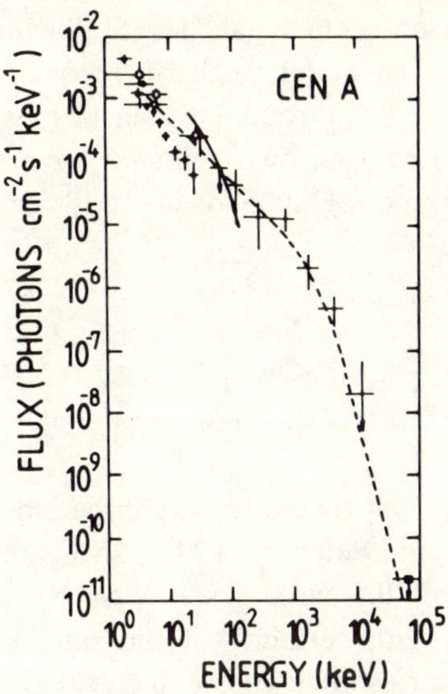

Figure 4.42. Summary of hard X-ray and γ-ray measurements for the radio galaxy Centaurus A (NGC 5128), after the summary by Fichtel (1982). The steepening in the spectrum above a few MeV is very marked in the case of all the extragalactic 'sources' of Figures 4.39, 4.40 and 4.41.

from the prominent Seyfert NGC 4151 up to about 10 MeV (see Dean and Ramsden 1981 for a useful summary), and the results are shown in Figure 4.41. No finite flux was claimed by the SAS II or COS B workers, but Houston and Wolfendale (1982) think that there might be a weak signal in the SAS II results ($E > 35$ MeV). Whether this is true or not the spectrum from this object, too, shows a steepening above an MeV or so. A similar situation exists for the Seyfert galaxy MCG 8-11-11 in that the galaxy has been detected up to a few MeV (Baker *et al.* 1981).

Radio galaxies would also be expected to be γ-ray emitters in that the electrons which generate the observed synchrotron radiation can also interact with gas to provide bremsstrahlung γ-rays. The prominent radio galaxies Cygnus A and Cen A are strong candidates, and γ-rays have indeed been detected from the latter, but only up to about 10 MeV (Figure 4.42) (although there is some evidence for its detection above 10^{11} eV and this is considered later). Including an upper limit in the SASII/COS B energy region there is again a spectral steepening above 1 MeV.

The final group of EG objects comprises 'normal' galaxies. Following Houston and Wolfendale (1982), the simplest model which can be used to make predictions is where γ-ray emission is proportional to galactic mass (for galaxies not grossly dissimilar in type to our own). In such a case, adopting a rate of emission of γ-rays from our Galaxy of $1.3 \times 10^{42}\ s^{-1}$ above 100 MeV (Worrall 1977), we have

$$F_\gamma = \frac{0.14}{4\pi} \frac{M}{M_G} \frac{1}{d_k^2}\ \mathrm{cm^{-2}\ s^{-1}} \tag{4.7}$$

where M/M_G is the mass of the galaxy in terms of the mass of our Galaxy and d_k is the distance in kpc.

Inspection of astronomical data shows that there are only three 'normal' galaxies which have a chance of being detected: LMC, SMC and M31; for these the expected fluxes are 3.7, 0.5 and $0.7 \times 10^{-7}\ \mathrm{cm^{-2}\ s^{-1}}$, respectively. None has been detected with certainty at the satellite energies but Riley (1982, private communication) has presented evidence for a flux of order $7 \times 10^{-7}\ \mathrm{cm^{-2}\ s^{-1}}$ for the LMC, to be compared with a more sophisticated model predicting $9 \times 10^{-7}\ \mathrm{cm^{-2}\ s^{-1}}$ for this galaxy (see also Houston *et al.* 1983). At least there is no gross discrepancy, but clearly this aspect of the subject is still in its infancy.

Andromeda (M31) has not yet been observed at the energies of interest here but, as will be demonstrated in Chapter 5, there is evidence for its detection at TeV energies.

4.7.3 The diffuse extragalactic flux

The evidence for a diffuse EG flux is based on the SAS II results (Figures 4.6 and 4.27) which indicate high latitude fluxes over and above the expected contribution from Galactic cosmic ray interactions in the ISM. The paucity of the data does not allow a positive identification, however; for example, inverse Compton interactions of electrons with the 2.7 K radiation in an extended Galactic halo could mimic an EG component and the present lack of statistics means that we cannot detect the expected non-isotropic distribution from Compton emission. However, we proceed and assume that the quanta *are* extragalactic.

At lower energies (X-rays), there is no doubt that there is an EG flux, and indeed its magnitude is considerable. The γ-ray spectrum fits well to the X-ray results, and this feature tends to confirm the EG origin of the γ-rays. Figure 4.43 shows the situation.

The most interesting feature seen is the apparent presence of a 'bump' in the region of a few MeV and this has led to many speculations. The

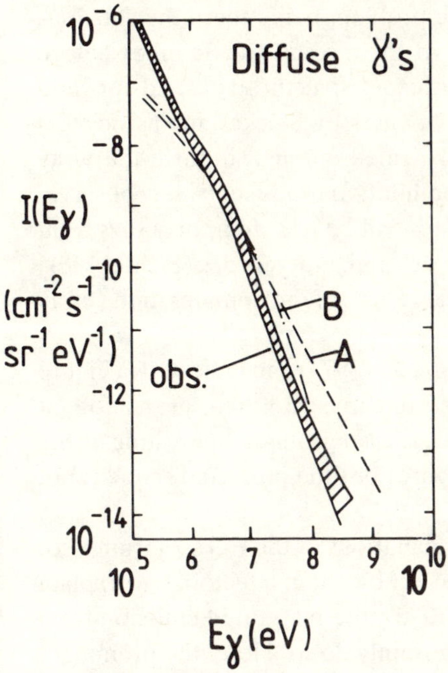

Figure 4.43. The energy spectrum of diffuse X- and γ-rays derived from the work of Trombka *et al.* (1977) and Fichtel *et al.* (1978). The 'bump' at a few MeV is apparent. There are many possible explanations; variants of a model due to Strong *et al.* (1974), and referred to in the text, are indicated.

bump has had a chequered history; it originated in Apollo satellite observations and was initially of considerably greater magnitude than shown, but much was shown later to be due to induced radioactivity in the detector. The spectrum shown in Figure 4.43 is, hopefully, the asymptotic limit.

A number of rather exotic suggestions have been put forward to explain the bump. Ginzburg (1968) and Stecker (1969) have independently pointed out that the enhancement could arise from π^0 decays in the Universe at early epochs when both the cosmic ray intensity and the gas density might have been bigger, the resulting γ-rays being red-shifted. Stecker (1969) has gone further and postulated that the interactions responsible were not pp (i.e. $p + p \rightarrow \pi^0 + \ldots$) but $p\bar{p}$, the antiprotons being primordial. The question of whether or not the Universe is baryon-symmetric, and EG $\bar{p}$ thereby being common, is still not resolved. The original model of Omnes (1969) based on a baryon-symmetric cosmology has had considerable opposition, the problem being the rather obvious one of how p and p̄ are separated in their respective families before annihilation occurs. Some form of domain structure is called for, and Brown and Stecker (1979) (and later papers by Stecker and associates) still consider that symmetry cannot be ruled out and, in turn, the γ-ray bump *could* be an indicator of the annihilation products. A key observation for the next generation of satellites will be to search for γ-rays from the interface between clusters of galaxies and anti-galaxies (Fichtel 1981, 1983), these units being essentially the results of the domains in the early Universe.

Another explanation, due to Strong *et al.* (1974) and indicated in Figure 4.43, is in terms of an EG origin of cosmic rays and their interaction, at early epochs, with the 2.7 K microwave background. Interestingly, the energetics are reasonable (see the figure) but the predicted shape is not too good.

There are other more mundane explanations of the EG spectrum also. Although normal galaxies are insufficient by about a factor 20 to explain the observed flux (e.g. Strong *et al.* 1976), the integrated emission from more energetic galaxies can almost certainly do so. Here, the prominent feature of a turn-down at a few MeV for 3C273 and NGC4151 referred to in the previous section adds weight to this suggestion. The energetics of the situation are also acceptable, although it will be appreciated that the paucity of detected sources causes considerable uncertainty.

Quite recently, yet another possible explanation for at least a part of the diffuse EG flux has been put forward. Strong and Bignami (1983) have claimed the detection of γ-rays from the direction of NGC 1275 in the Perseus cluster (the signal can be seen rather clearly in Figure 4.29). This cluster is at a distance of 116 Mpc (using a Hubble constant $H_0 = 50$ km s^{-1} Mpc^{-1}) and the flux claimed is 8×10^{-7} cm^{-2} s^{-1} above 70 MeV. The manner in which the γ-rays are produced is not clear (one or more very energetic galaxies in the cluster or cosmic rays interacting with the intercluster gas, or both?) but they do seem to be present. Houston *et al.* (1984) have followed this up by examining the SAS II data base in detail in a search for an excess flux from the directions of specified clusters. Although the statistical precision is understandably poor, there does seem to be an excess, and when integrated over all clusters, suitably weighted according to γ-ray 'strength', the predicted contribution to the EG diffuse flux is $(6.0 \pm 2.4) \times 10^{-4}$ cm^{-2} s^{-1} above 35 MeV, to be compared with the observed value: $\approx 7.4 \times 10^{-4}$ cm^{-2} s^{-1}. Here, then, is another contender.

A final brief comment can be made about the role of the EG γ-ray flux in constraining models of cosmic ray origin. If cosmic rays were of EG origin and had the same intensity everywhere in the Universe, or were concentrated in galaxy clusters, then their interactions with EG gas would generate a higher γ-ray flux than is seen (Said *et al.* 1982). In fact, the excess of prediction over observation is not large for the constant cosmic ray intensity case but this situation is unlikely; more likely is the model in which cosmic rays are generated in active galaxies in clusters, and here the excess of prediction over observation is large. The argument against EG cosmic rays is quite a strong one.

4.8 New experimental programmes

4.8.1 General remarks

Although balloon flights are still continuing from time to time, these are devoted mainly to rather low γ-ray energies ($\lesssim$10 MeV) and their necessarily short durations yield rather poor quality data. It is from new satellite experiments that we expect major new advances to occur and here there are two projects in hand: GAMMA I and the GRO.

4.8.2 GAMMA I

Construction of the equipment for this 'observatory' is nearing completion and there should be an early launch (1986) using a Soviet satellite. The detector is similar to those on SAS II and COS B, namely a multilayer spark chamber with anticoincidence protection against charged cosmic ray events. Improvements over the earlier satellite detectors include a more accurate energy calorimeter, and a sensitive area higher by nearly a factor three (although the product of area times solid angle is about the same). Most important, the accuracy of determination of γ-ray directions will be greater.

As with COS B, the project is international in character, the contributors being the USSR (Moscow Space Research Institute, Moscow Physical Engineer Training Institute, Lebedev Physics Institute of Moscow, and A. I. Ioffe Physical Technical Institute of Leningrad); France (Section d'Astrophysique du CEA/Saclay, and Centre d'Etude Spatiale des Rayonnements); and Italy (Istituto di Fisica Cosmica of Milano, Istituto di Astroficia Spaziale at Frascati, and Istituto TESRE at Bologna).

4.8.3 Gamma Ray Observatory

Due for launch in 1988 is NASA's Gamma Ray Observatory (GRO), a project which should provide considerable data for analysis over the ensuing decade.

There will be four instruments covering between them the energy range 30 keV–30 GeV.

Following the description given by Fichtel (1982), their capabilities can be listed as follows:

(i) A survey of high energy γ-ray sources and diffuse emission with a point source sensitivity of better than 10^{-7} cm^{-2} s^{-1}, angular resolution of $\approx$0.1 for strong sources and an energy resolution of $\approx$15 % at energies above 100 MeV.

(ii) A survey of sources and diffuse emission with sensitivities $\approx 10^{-5}$ cm^{-2} s^{-1} and energy resolution around 10 % between 0.1 and 30 MeV.

(iii) Detection and identification of nuclear γ-lines with an energy resolution of $\approx$4 % and sensitivity $\approx 5 \times 10^{-5}$ cm^{-2} s^{-1}.

(iv) Observations of γ-ray bursts, including studies of their spectral and temporal behaviour.

The planned orbit is near circular at 400 km and an inclination of 28.5°. Pointing directions will be known to an accuracy of about 2′ and absolute times will be accurate to about 100 μs.

The design characteristics of the four instruments can be summarised as follows:

Oriented scintillation spectrometer experiment (OSSE). Energy range 0.1–10 MeV. Detectors: four large actively shielded and passively collimated NaI scintillators. Field of view 5° × 11° FWHM. Contributing laboratories: Naval Research Laboratory; North-Western University; Royal Naval College, London; Rice University.

Compton telescope (COMPTEL). Energy range 1–30 MeV. Detectors: upper liquid scintillator, lower NaI (Tl) crystal – the γ-ray interacts to produce Compton electrons in both detectors. Field of view $\approx$1 sr, resolution 'a few degrees'. The instrument can determine the polarisation of the radiation. Contributing laboratories: Max Planck Institüt für Extraterrestrische Physik; University of Leiden; University of New Hampshire; European Space Agency, ESTEC.

Energetic gamma-ray experiment telescope (EGRET). Energy range 20 MeV–30 GeV. Detectors: multilevel thin plate spark chamber system (similar to SAS II and COS B), NaI (Tl) total energy calorimeter beneath the spark chamber and plastic scintillator anticoincidence dome. Precision: for sources of moderate strength the position should be determined to about 10 minutes of arc, for diffuse emission spatial variations in the energy spectrum should be measurable on a scale of a few degrees. Contributing laboratories: Goddard Space Flight Centre; Stanford University; Max Planck Institüt für Extraterrestrische Physik; Grumman Aerospace Corporation.

Burst and transient source experiment (BATSE). Energy range: 50–600 keV. Detectors: eight wide field modules; four have the same viewing path as the other telescopes on GRO and four view the opposite hemisphere. Time resolution $\approx 100\ \mu s$, burst location accuracy $\approx 1°$, sensitivity $\approx 6 \times 10^{-8}$ erg cm^{-2} for a 10 s burst. Laboratory involved: Marshall Space Flight Centre.

5

ULTRA-HIGH ENERGY GAMMA-RAYS

5.1 Introduction

Studies of ultra-high energy γ-rays (UHEGR)*, i.e. γ-rays at energies greater than 100 GeV, have an important bearing on the problem of the origin of the cosmic radiation. There is, as yet, no universally accepted identification of either the sources or the mechanisms of production of cosmic rays, though, as was pointed out in Chapter 4, there are strong arguments made in favour of some. The problem is confounded by the fact that cosmic rays, almost all of which are charged particles, undergo frequent deflections in the interstellar magnetic fields, making it impossible to know the source directions. Thus, even a charged primary cosmic ray particle of energy as high as 10^{15} eV has a Larmor radius in the ISM of only $\sim$0.3 pc and has its initial direction almost isotropised. Electrically neutral radiation is free from this problem. The more commonly occurring neutral particles are neutrons, neutrinos and γ-rays. Neutrons are unstable; they would not survive from source to earth even after allowing for relativistic time dilatation, at least at energies $\leqslant 10^{15}$ eV and for source distances >100 ly. Neutrinos, being weakly interacting, are not easy to detect. γ-rays, on the other hand, are ideal in view of both their production and interaction cross sections being rather high and their being stable.

There are strong indications that ultra-high energy charged particles (specifically electrons) exist in regions far removed from the solar system.

* Gamma-rays in the range 100–10000 GeV are sometimes referred to as 'very-high energy gamma-rays', whilst those above 10^{14} eV are termed 'ultra-high energy gamma-rays'. Here, for the sake of brevity, we use UHEGR for both.

Shklovskii (1953) suggested that the optical continuum from the CRAB Nebula must be due to synchrotron emission by high energy electrons and positrons in the nebular magnetic fields, and subsequent observations on the polarisation of photons at radio, optical and X-ray wavelengths confirmed this suggestion. This shows that relativistic electrons of energy up to at least 10^{14} eV exist in the CRAB nebula, a fact which has made the prospect of detecting UHEGR from this and other similar celestial sources very attractive. If a finite flux of UHEGR is detected, the source direction is known immediately, being part of the discovery process itself. Further studies of the energy spectra and the time variations of intensity are likely to throw light on both the physical conditions and the charged particle acceleration mechanisms operating in the environs of the source object.

Production mechanisms of UHEGR have been reviewed in the past by several authors; see, for example, Felten and Morrison (1966), Fazio (1967) and Stecker (1971). As mentioned in Chapter 1, UHEGR can be produced in the interactions of high energy protons with matter ($p + p \rightarrow \pi^0 \rightarrow 2\gamma$) and of high energy electrons with matter (bremsstrahlung), with magnetic fields (synchrotron and curvature radiation mechanisms) and with ambient radiation fields (inverse Compton process).

Once produced, UHEGR travel from the source to the observer undeflected, although some attenuation by γ–γ collisions may occur en route. The absorption mechanism was considered in Chapter 1. It was shown there that the threshold for γ–γ collisions is given by

$$E_\gamma > m_e^2 c^4/\varepsilon \tag{5.1}$$

where m_e is the rest mass of electron and ε is the energy of the ambient photon. The consequence is that extragalactic space, in which ε relates to the 2.7 K radiation, is transparent out to about 100 Mpc for $E < 10^{14}$ eV and ~ 10 kpc for $E_\gamma \sim 1.10^{15}$ eV.

Another mechanism for absorption, but in a very different environment, is pair production by interaction with intense magnetic fields in the magnetospheres of pulsars. The threshold photon energy for this process (Sturrock 1971, Manchester and Taylor 1977) is given by

$$E_\gamma(\text{eV}) > 4 \times 10^{18}/B_\perp(\text{G}) \tag{5.2}$$

where $B_\perp$ is the magnetic field transverse to the direction of the γ-ray.

Experiments on UHEGR from celestial bodies are beset with the difficulty of having to detect very low fluxes in the presence of considerable backgrounds from charged cosmic rays. The rarity of UHEGR can be gauged from the fact that the total number of UHEGR claimed to have been detected during the past 25 years in approximately as many experiments is only a few thousand. To date there have been about ten claims made by experimenters for having detected UHEGR from celestial sources at the 4–5σ level over the energy range 100 GeV $< E_\gamma <$ 10^4 TeV and much more work needs to be carried out before this branch of γ-ray astronomy becomes as well established as the others.

If the energy spectrum observed at medium high energies (e.g. from the CRAB pulsar) is extrapolated to E_γ = 1000 GeV it is found that the expected number of γ-rays incident over 1 m^2 area is only about 16 y^{-1}. This number is so low as to make it almost impossible here to use balloon- or satellite-borne detectors. Fortunately, however, observations on UHEGR can be carried out by ground-based experiments by exploiting the transducing role of the terrestrial atmosphere itself: an incoming UHEGR interacts with the atmosphere, producing a very broad distribution of secondaries, tertiaries etc., and it is these that a ground-based detector can record. The γ-ray shower axis can be as far away as 100 to 200 m from the detector and yet the detector would be able to detect the progeny. This feature allows one to use detectors of modest physical size (a few to a few tens of square metres) and be able to record the incidence of celestial γ-rays over tens of thousands of square metres. This gain in the effective collection area by more than three orders of magnitude is what makes it possible for both the atmospheric Cerenkov and extensive air shower techniques to detect the very small flux of cosmic UHEGR.

Extensive air showers initiated by charged cosmic rays constitute a large irreducible background to the detection of UHEGR. Further, since cosmic rays arrive uniformly in time (albeit with statistical fluctuations), and almost isotropically in space, present experiments cannot isolate the diffuse component of UHEGR. The aim, therefore, is to detect UHEGR from point sources which stand out as spikes in an otherwise uniform spatial distribution. If, in addition, the UHEGR intensity varies with time in a well defined fashion (such as happens for pulsars) it provides an additional handle to the experimenter.

Basically two techniques are employed in the detection of UHEGR, namely (i) the atmospheric Cerenkov technique in the energy range 100 GeV $< E_\gamma <$ 100 TeV, and (ii) the extensive air shower technique at $E_\gamma >$ 100 TeV. The two techniques and various experimental results are described in Sections 5.2 and 5.3, respectively. The implications of the results and various theoretical models of the UHEGR sources are presented in Section 5.4, and finally a summary is given in Section 5.5.

5.2 Experiments at 100 GeV < E_γ < 100 TeV

5.2.1 The atmospheric Cerenkov technique

Blackett (1948) first suggested that Cerenkov light produced by charged cosmic rays in the terrestrial atmosphere should contribute about 10^{-4} of the night sky optical background. Galbraith and Jelley (1953, 1955) demonstrated that cosmic ray showers do indeed produce Cerenkov light pulses in the atmosphere. Chudakov *et al.* (1962) were the first to apply this technique to the detection of celestial γ-rays. (See Jelley and Porter (1963) and Jelley (1967) for earlier reviews, and Porter and Weekes (1978) and Ramana Murthy (1980) for the more recent. See also the Proceedings of the International Workshop on Very High Energy Gamma Ray Astronomy, Ootacamund, India, 1982, edited by P. V. Ramana Murthy and T. C. Weekes.)

A high energy γ-ray, on entering the earth's atmosphere, initiates an electromagnetic cascade shower. Electrons and positrons in the shower with velocities above the local light velocity (c/μ) produce Cerenkov light, μ being the refractive index in air in the region where the light is produced. The dependence of energy threshold, emission angle and the rate of production of Cerenkov radiation on the depth in the atmosphere is shown in Figure 5.1. Although the electrons and photons in the cascade are absorbed in the atmosphere, the optical Cerenkov light penetrates the atmosphere, reaching the detectors at mountain altitudes or at sea-level. The Cerenkov light flash (typical duration $\Delta T \approx 10$ ns), created by a 300 GeV γ-ray, for example, may contain a few million photons spread over distances of the order of several hundred metres from the axis. If the field of view of a detector is restricted to a cone of 1° in diameter, the Cerenkov photon density is $\sim 7\ \mathrm{m}^{-2}$ near the core of the shower, dropping off drastically beyond about 100 m from the axis. To a good approximation, the Cerenkov photon density is proportional to the γ-ray energy. Several authors have calculated the lateral distribution of Cerenkov photons; see Browning and Turver (1977), for example. The calculations (see Figure 5.2) reveal that the photon density depends on the half-angle of the cone of view and is affected by the geomagnetic field. The authors have also shown that, for the same γ-ray energy, the magnitude and shape of the lateral distribution of the Cerenkov photon density varies from shower to shower due to fluctuations in shower development. Since the magnitude of the Cerenkov light is very small, experiments can be carried out only during clear moonless nights, and this makes the duty factor very small, typically 10–20 %.

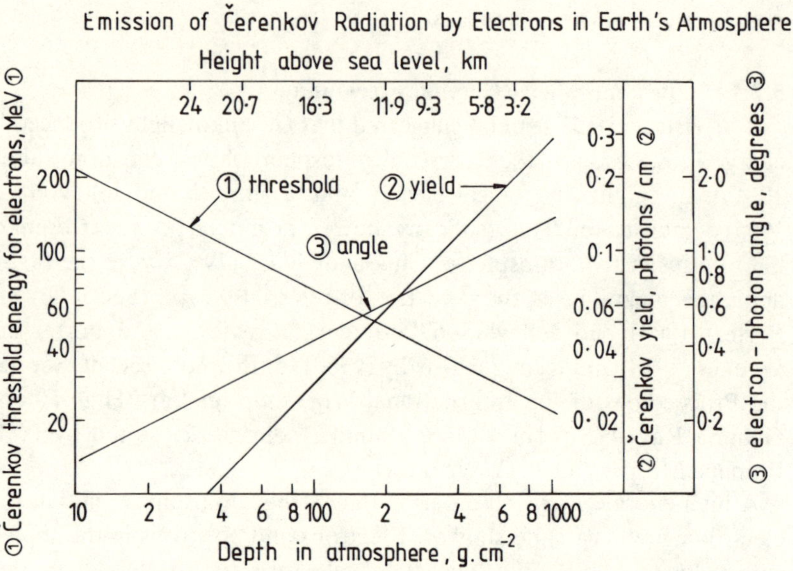

Figure 5.1. Variation with depth in the terrestrial atmosphere of (1) threshold energy for electrons to emit Cerenkov radiation, (2) intensity of emission and (3) angle of emission of Cerenkov radiation. For electromagnetic cascades the depth region 100–400 g cm^{-2} in the atmosphere is the most relevant.

A celestial source may appear to be a point as seen from earth, but Coulomb scattering of the electrons in the cascade and the finite emission angle of Cerenkov light spoil the resolution and make the source look like a diffuse object of 1° to 2° in diameter. In practice, the low intensity Cerenkov light is collected and focused by a large parabolic mirror on to the cathode of a fast photomultiplier, in front of which is a mask to limit the view to a cone of 1° to 3° in diameter. It is routinely demonstrated in each experiment (see, for example, Gupta 1983) that most of the Cerenkov photons arrive in a flash of less than 10 ns in width and with mutual angular separations of $<1.5°$.

The night sky background, NSB, is $\approx 7 \times 10^7$ photons cm^{-2} s^{-1} sr^{-1} when the view does not include a bright star (Ramana Murthy and Tonwar 1976). NSB and the atmospheric Cerenkov radiation produced by extensive air showers (EAS) initiated by charged primary cosmic rays constitute the background to the γ-ray signal. NSB is almost completely eliminated by taking fast coincidences ($\tau \sim 10$ ns) between several photomultipliers looking at different mirrors. NSB can also be eliminated by

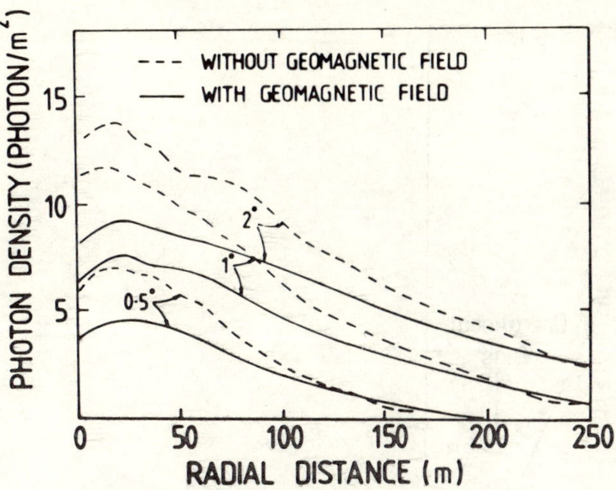

Figure 5.2. Lateral distributions found by Browning and Turver (1977) of Cerenkov photons produced by $e^{\pm}$ in the atmosphere with and without geomagnetic field for an incident γ-ray energy of 100 GeV. Calculations apply for an altitude of 2300 m above sea-level. Numbers attached to curves represent the half-angle of the field of view in degrees. The figure is taken from Porter and Weekes (1978).

demanding a majority logic among several photomultipliers or by adding pulses from all the photomultipliers and then biasing off at a suitable discrimination level; see Figure 5.3. With NSB almost eliminated, cosmic ray-initiated EAS remain the only background to the UHEGR signals, but these are not easy to remove. As it is, the cosmic ray background primaries are less efficient in producing a trigger than the UHEGR signal, for two reasons: (i) at energies $>10^{12}$ eV, much of the energy in the former case is 'wasted' in producing nuclear excitations and in the subsequent decays of pions and muons into neutrinos. This energy is thus lost to the electromagnetic component responsible for producing the Cerenkov emission, and (ii) at energies below 10^{12} eV, the nucleus-induced EAS become not only even more inefficient for the reason just stated but the secondary pions are emitted at larger angles and thus spread out the generated Cerenkov light to larger radial distances. The progressively increasing advantage that γ-rays enjoy over cosmic rays with decreasing energy is clearly demonstrated in Figure 5.4, taken from Turver and Weekes (1978).

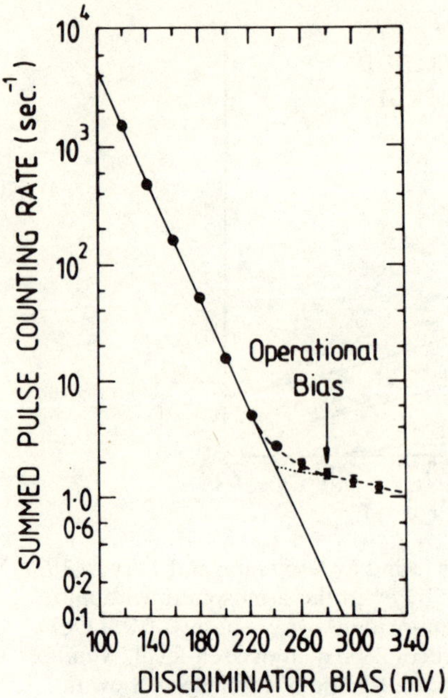

Figure 5.3. Signals from all the 18 photomultipliers, each viewing a parabolic mirror in the Ooty group's set up, are summed and discriminated. The figure shows the counting rate as a function of discriminator bias. While the steep portion represents contributions from photomultiplier noise and NSB, the flatter portion is due to air showers and γ-ray showers. If the discriminator is operated at 280 mV bias, there will be a negligible number of chance coincidences among the triggers for showers (Gupta 1983).

Grindlay (1971) made use of the Cerenkov light produced by *muons* in EAS (the 'second maximum') as a veto to reduce the cosmic ray background, but this technique is efficient only at $E \gtrsim 10^{13}$ eV because at lower energies there are too few muons in the showers. It is unfortunate that there are no standard candles (UHEGR sources) in the sky that can be used to calibrate the detector systems, but, though troublesome as a background, the cosmic ray-produced Cerenkov radiation has some value for calibration, the charged primary cosmic ray energy spectrum being well studied in this energy range.

The cosmic ray background rate is proportional to the product of collection area, A, and solid angle of view, $\pi\theta^2$ (θ small), while the γ-ray (signal) rate from a discrete source is proportional to A. If the experiment operates for a time T, then the standard deviation in the background counts is $\propto (AT\theta^2)^{1/2}$, while the excess counts due to the signal are $\propto AT$.

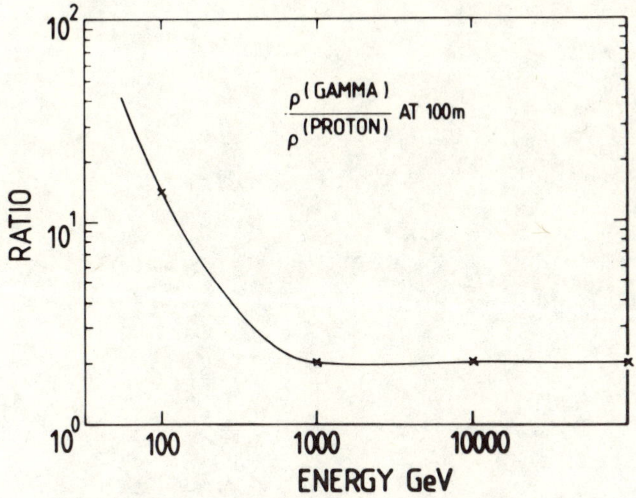

Figure 5.4. The ratio of Cerenkov photon densities ρ(GAMMA)/ρ(PROTON), at 100 m from axis (for γ- and proton initiated cascades), is shown as a function of energy of the initiating particle; taken from Turver and Weekes (1978).

The signal-to-noise ratio, then, is proportional to $AT/(AT\theta^2)^{1/2} = A^{1/2}T^{1/2}\theta^{-1}$. To sharpen the γ-ray signal, one has, therefore, to maximise the observation time, T, and decrease θ, the half angle of the cone of view. There is a limit, however, of $\theta \sim 0.5°$, below which one cannot decrease the angle because the fraction of Cerenkov light collected, and consequently the detection efficiency of the experiment, falls very rapidly below this value. To maximise the collection area for a given γ-ray energy, one decreases the Cerenkov photon density threshold by using larger mirror area. This also results in the more important advantage of lowering the γ-ray energy threshold.

If the source is pulsating, the sensitivity of a given detector system is improved by a factor of 5 to 10 compared to that for a steady source. However, searches for pulsed γ-rays can be significant only if one has an accurate knowledge of the period, P; in the case of a pulsar, where P is varying with time, one needs also the period-derivative, $\dot{P}$, of the object, this being determined from measurements carried out at other frequencies at an epoch as close to that of the γ-ray experiment as possible.

In the past, groups from the USSR, UK, Ireland, USA, Australia and India have detected UHEGR using the Cerenkov technique; see the review by Porter and Weekes (1978) for details of some of these experi-

Figure 5.5. A picture of the 10 m reflecting night sky Cerenkov telescope of the Fred Lawrence Whipple Observatory at Mt. Hopkins, USA. (Smithsonian Institution – courtesy of Prof. T. C. Weekes.)

ments. We will mention here only a few of them to illustrate the detector systems generally used.

The 10 m diameter dish of the Smithsonian Astrophysical Observatory at Mt. Hopkins, USA, is shown in Figure 5.5. It consists of 248 individual mirror elements and has a computer-controlled alt-azimuth mounting (Cawley *et al*. 1982). The dish has a focal length of 7.3 m and an angular resolution of 0.35°. In its latest mode of operation, the dish is equipped with 37 photomultipliers arranged along its focal plane.

Figure 5.6 shows the 18 mirror array of the Tata Institute of Fundamental Research. The array, consisting of eight parabolic mirrors of 1.5 m diameter and ten of 0.9 m diameter (Bhat *et al*. 1980), is operated at Ooty, India. Each mirror is equatorially mounted and is viewed by a photomultiplier placed at its focus. During most runs, the mirrors are grouped into four banks, photomultiplier pulses are summed and discriminated in each bank and a majority logic (any three out of four banks) is used to eliminate NSB and to trigger on UHEGR and cosmic ray EAS.

Figure 5.6. The 18 parabolic mirror array to detect Cerenkov light produced by showers. The array was operated by the Tata Institute of Fundamental Research group at Ootacamund (Ooty), India. Two small mirrors were not used in the experiment. (Photograph by A. R. Apte.)

The University of Durham (UK) has operated a spaced array of four γ-ray telescopes at Dugway, Utah, USA. Each telescope consisted of three paraxial parabolic mirrors of 1.5 m diameter mounted on a computer-steered alt-azimuth platform (Gibson *et al.* 1982*a*, *b*). The telescopes were deployed one each at the centre and apices of an equilateral triangle of side 100 m. Each telescope acted independently and generated its own trigger by a three-fold coincidence between the photomultipliers viewing the three mirrors.

The objective is usually to study if a celestial object is steadily emitting UHEGR, and, to do this, a common practice is to aim the mirrors at a point in the sky in the path of the object and hold them fixed there. The event rate is expected to change significantly as the object moves in and out of the detector field of view (drift scan mode). A variant of this technique, the 'tracking mode', is to track the object for some time between two spatial angles θ_1 and θ_2, followed by tracking the now empty field for the same time between the same angles and then compare the resulting 'on-source' and 'off-source' rates.

For pulsed emission studies, on the other hand, the mirrors are locked to the object under investigation and made to track it for as long as possible (typically 1–6 h); event times and pulse heights are recorded. Later, the event times are examined for the expected periodicity.

5.2.2 Results on steady emission

Porter and Weekes (1978) have summarised UHEGR flux values and upper limits in the energy range $10^{11} < E_\gamma < 10^{14}$ eV from nearly a hundred celestial objects looked at by various experiments. The objects included pulsars (both steady and pulsed emissions), novae, SN, SNR, magnetic variables, flare stars, X-ray sources, BL Lac type objects, galaxies, galactic clusters and quasars. (See also Grindlay 1982 for another useful survey of some of the results.)

5.2.2A CRAB nebula/CRAB pulsar

The CRAB nebula, the best known SNR in the Galaxy, is located about 2 kpc away from the solar system. The SN explosion took place in the year 1054 AD and is well documented in ancient Chinese records. Its angular size is a few arc minutes as seen from the Earth. A 33 ms pulsar was discovered in the remnant by Staelin and Reifenstein (1968) and Comella *et al.* (1969). The pulsar is known to emit at other frequencies besides radio: optical (Cocke *et al.* 1969), X-ray (Fritz *et al.* 1969) and γ-ray (Hillier *et al.* 1970). With such ample evidence for the existence of high energy particles and the associated generation processes, it was natural that the CRAB nebula and pulsar should be favourite objects to look at in almost all the previous experiments on the UHEGR.

Chudakov *et al.* (1962, 1965) carried out the first extended observations on the CRAB and set an upper limit of 5×10^{-11} photons cm^{-2} s^{-1} at $E_\gamma > 5 \times 10^{12}$ eV. Fruin *et al.* (1964) and Long *et al.* (1965) set upper limits of 1.3×10^{-11} photons cm^{-2} s^{-1} at $E_\gamma > 2.7 \times 10^{13}$ eV. Fazio *et al.* (1972), from a three year observation on the CRAB nebula with the 10 m dish at Mt. Hopkins, found a positive effect at the 3.1σ level. The flux, averaged over all the observation time, was given as $(4.4 \pm 1.4) \times 10^{-11}$ photons cm^{-2} s^{-1} at $E_\gamma > 2.5 \times 10^{11}$ eV. The authors noted, however, that much of the effect was contributed by runs during three separate periods, each between 60 and 120 days after an observed spin-up (glitch) of the pulsar; see Figure 5.7. If averaged only over these periods, the flux went up to $(1.21 \pm 0.24) \times 10^{-10}$ photons cm^{-2} s^{-1}, a 5σ effect.

This kind of situation, of detecting a signal at the 3–5σ level, but not finding it every time, is not confined to this experiment or to the CRAB

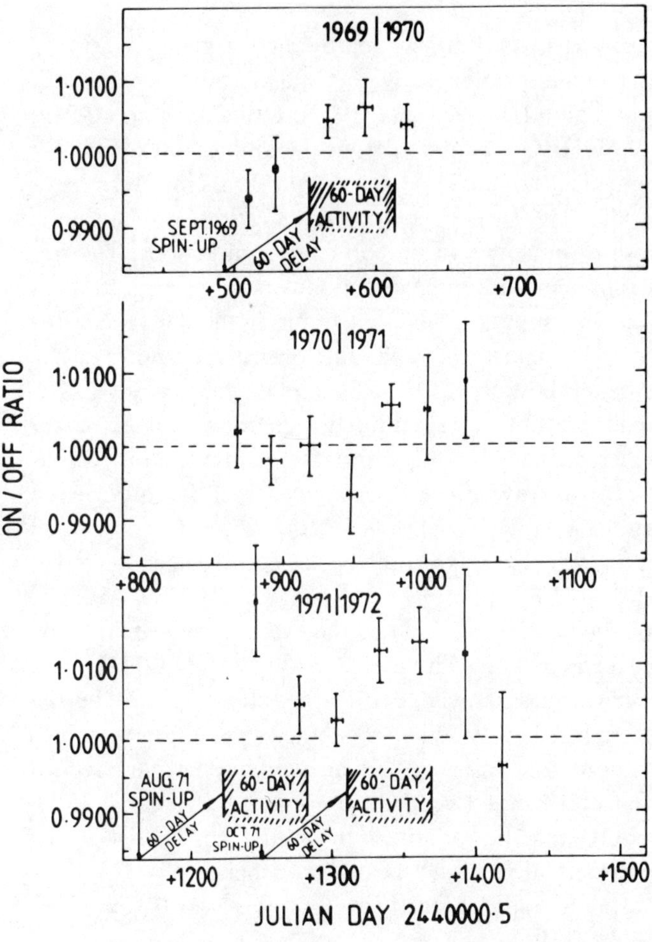

Figure 5.7. Ratio of the counts recorded from the CRAB nebula/pulsar to those from an arbitrary reference direction (Fazio *et al.* 1972). The authors operated the Mt. Hopkins 10 m reflector for three consecutive years. Notice the relationship in time between the pulsar spin-up epoch and on/off ratios greater than unity, signifying UHEGR emission by the CRAB nebula/pulsar.

nebula. The theme recurs frequently, as will be seen in the following pages. On the basis of the time-averaged flux detected by Fazio *et al.* (1972), the luminosity for a steady emission of UHEGR from the CRAB nebula/pulsar is ~2×10^{34} erg s^{-1}. Inferences that can be drawn from these results will be outlined later, in Section 5.4.

5.2.2B Cen A

Centaurus A (NGC 5128), a 'nearby' active galaxy (distance ~6 Mpc) is known to exhibit violent activity at radio frequencies. It is also visible in the X-ray region (Bowyer *et al.* 1970). Grindlay *et al.* (1975*a*) looked for possible UHEGR emission from this object during 1972–4 using the pair of 7 m dishes of the Hanbury Brown–Twiss interferometer at Narrabri, Australia. The authors compared the rates from the source direction with those from neighbouring 'off-source' directions. The frequency of excess counts in the source direction versus the magnitude of excess counts in units of standard deviations, during the data cycles for three years, is shown in Figure 5.8. While the lower panel shows all the events, the upper panel shows events which are mainly due to UHEGR. This selection is made possible by rejecting showers showing a secondary Cerenkov pulse in a beam slightly offset from the main one and attributable to muons in a cosmic ray initiated EAS. In each of the three years, the authors detected a positive signal (3.9, 1.0 and 2.5σ). The combined signal, amounting to a 4.6σ excess over the background, corresponded to a flux of $(4.4 \pm 1.0) \times 10^{-11}$ photons cm^{-2} s^{-1} at $E_\gamma > 300$ GeV. The authors give a probability of $<2 \times 10^{-5}$ that the excess counts are not due to a γ-ray source. The beam width being only 0.45° (FWHM), the extended (~5°) radio lobes associated with this object cannot be the source of the detected UHEGR. When combined with X-ray measurements, the observations are consistent with an energy spectrum having a differential spectral index of -1.7. The experiment by Grindlay *et al.* suggests that the source may be variable in the emission of UHEGR in a similar manner to the situation at microwave frequencies.

If isotropic emission is assumed, the luminosity is $\sim 2 \times 10^{41}$ erg s^{-1} at 300 GeV $< E_\gamma <$ 3000 GeV.

5.2.2C M31

M31, the Andromeda nebula, belongs to the local group of galaxies and is located at a distance of 670 kpc from the solar system. Though Weekes *et al.* (1972) had previously set an upper limit of 3.8×10^{-10} photons cm^{-2} s^{-1} at $E_\gamma > 1.4 \times 10^{11}$ eV from this object, Dowthwaite *et al.* (1984*a*) have now found a 3σ positive signal at a flux level of 2.2×10^{-10} photons cm^{-2} s^{-1} at $E_\gamma > 10^{12}$ eV. The result is based on drift scans made during 1983 at Dugway, Utah, USA. There were 240 ± 79 excess counts in the source direction. The null (no signal) hypothesis fits the data only at a level of 1 % likelihood. According to the authors, the luminosity of M31 at $E_\gamma > 1000$ GeV is 4×10^{40} erg s^{-1},

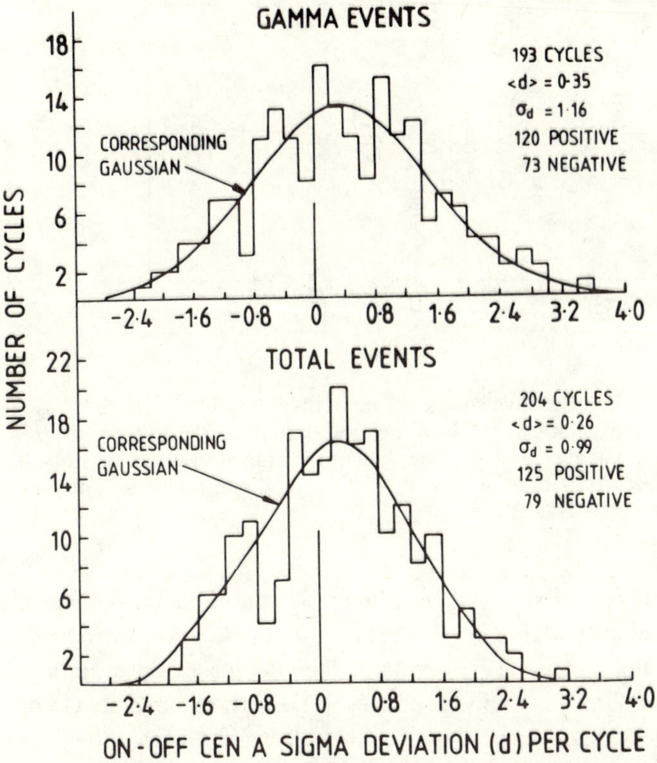

Figure 5.8. Distributions of observed deviations, in standard deviations, from the background in the direction of Cen A in the experiment of Grindlay *et al.* (1975*a*). All data (1972–4) are included. The mean number of standard deviations, $\langle d \rangle$, per cycle and the standard deviations of the histograms, σ_D, are shown. Lower panel: distribution for all showers; upper panel: showers enriched in γ-rays by rejecting some which showed signals from muons. In both panels there are more positive deviations than negative signifying UHEGR emission by Cen A.

assuming isotropic emission and a power law differential spectrum with a spectral index of −3.

5.2.2D Galactic Plane

The Crimean Astrophysical Observatory group (Fomin *et al.* 1977, Stepanian *et al.* 1982) find that there is a deficit $-(0.7 \pm 0.2)$ % of high energy events from the Galactic Plane over a wide range of longitudes: $320° < l < 170°$, $|b| < 1.5°$. The result is shown in Figure 5.9. Though it is only a 3.5σ effect, it is significant that a similar deficit was seen

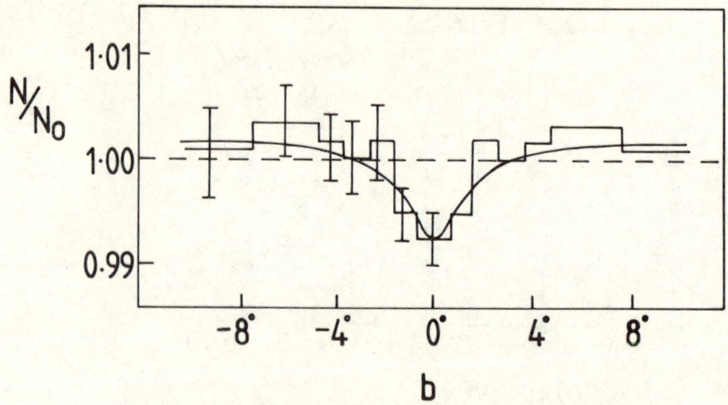

Figure 5.9. Relative counting rate of the EAS $N/\langle N\rangle$, where $\langle N\rangle$ is the average counting rate, as a function of Galactic latitude, b, after Stepanian *et al.* (1982). The data are averaged over a wide range of Galactic longitude, $320° < l < 170°$. The dip at $b = 0$ is of magnitude $-(0.7 \pm 0.2)\,\%$.

in several other experiments: by Chudakov *et al.* (1964), Weekes *et al.* (1972), Grindlay *et al.* (1975*b*), Weekes *et al.* (1979), and Dowthwaite *et al.* (1984*b*). Fomin *et al.* (1977) explained this decrease by postulating that the Galactic Plane actually produces UHEGR by inverse Compton scattering of starlight on very high energy electrons and that the dip seen in the UHEGR emission is due to the light radiation density at $|b| < 1.5°$ being lower owing to absorption by dust in the Galactic Plane. Varying NSB could be a problem here, however, leading to a possible decrease in the gains of the photomultiplier detectors, although some of the experiments that saw the effect provided constant illumination on photomultipliers by servo-controlled padding lamps, and thereby reduced the background problem. More work is desirable to consider the effect fully established, however.

5.2.2E SS 433

Gorham *et al.* (1982) reported seeing a 2.7σ signal from SS 433 (the 433rd member of a catalogue of peculiar stars compiled by Stephenson and Sanduleak 1977) during the period June 18–28, 1982, corresponding to the peak in the 164 d Doppler cycle of the source. The estimated flux is $\sim 10^{-10}$ photons cm^{-2} s^{-1} at $E_\gamma > 1$ TeV. The result is marginal and needs confirmation by an independent experiment.

Results presented above on steady emission of UHEGR by celestial sources are summarised in Table 5.1.

Table 5.1. *Results on steady emission of UHEGR*

Object	Group	Date	Effect (σ)	$E_{\gamma\,min}$ (TeV)	Flux ($10^{-11}\,cm^{-2}\,s^{-1}$)	Reference
CRAB nebula/pulsar	Lebedev	1962	—	5	<5	Chudakov *et al.* (1962, 1965)
	Dublin	1964–5	—	27	<1.3	Long *et al.* (1965)
	Mt. Hopkins	1969–72	3.1	0.25	4.4	Fazio *et al.* (1972)
			5.0*	0.25	12.1*	
	Ooty	1982	—	1	<20	Gupta *et al.* (1982) (3σ limit)
Cen A	Narrabri	1972–4	4.4	0.3	4.4	Grindlay *et al.* (1975*a*)
M31	Mt. Hopkins	1972	—	0.14	<38	Weekes *et al.* (1972)
	Durham	1983	3	1	22	Dowthwaite *et al.* (1984*a*)

* When averaged over selected periods only.

5.2.3 Results on pulsed emission

Pulsars are high on the list of potential objects worth observing by UHEGR experiments for a number of reasons: (i) Independent evidence exists that high energy processes are occurring in their vicinity. (ii) The precisely pulsed nature of emission relieves the experimenter from devoting as much (or even greater) amount of time to studying the cosmic ray background and systematic errors. (iii) If the data are divided into N bins in the phase plot, the cosmic ray background reduces by a factor N, as a result of which the sensitivity of an experiment goes up by $N^{1/2}$, a factor which is usually ~5–10.

The higher the spinning down energy loss rate ($\propto \dot{P}P^{-3}$), and the smaller the distance of a pulsar, the higher is the chance of detecting it. The figure of merit is given by

$$FM \propto \dot{P}/(D^2P^3), \text{ i.e. } \propto(\tau_{ch}D^2P^2)^{-1} \tag{5.3}$$

Here D is the distance to the pulsar, P its period and τ_{ch} the characteristic age of the pulsar. One therefore concentrates on the nearest, youngest and fastest pulsars. The recently discovered millisecond pulsars are an exception. Having been recycled and spun-up by accretion (Radhakrishnan and Srinivasan 1981), they are probably not as active as the first-born pulsars. Since the UHEGR signal is very weak, one cannot determine the pulsar elements (P, $\dot{P}$, epoch) from the data themselves; it is necessary to obtain the elements from observations made at other frequencies. Pulsars are known to undergo sudden speed-ups (glitches) and it is desirable, therefore, to obtain the elements from as recent an observation as possible.

Because the UHEGR signal is so weak it is necessary to accumulate data over long stretches of time ($T \sim 100$ d), at the same time keeping the phase information. If a pulsar of period P is observed for a time T, then the number of pulsar cycles elapsed, N, is given by $N = T/P$. In order to avoid a wash-out of the signal due to smearing, ΔN has to be kept small, $\lesssim 0.01$. This, then, calls for accuracies of $\sim 10^{-10}$ in time keeping ($\Delta T/T$) and in the period ($\Delta P/P$), in the case of the CRAB pulsar. For time keeping, one uses an atomic clock directly or, more generally, a laboratory quartz clock regularly calibrated with time signals derived from an atomic standard some distance away. The Doppler shifts, $\Delta P/P$, introduced in P by the Earth's orbital motion and rotation can be quite large, $\sim 10^{-4}$ and 10^{-6}, respectively; see Figure 5.10. To keep clear of these effects, the event arrival times are reduced to the solar system barycentre.

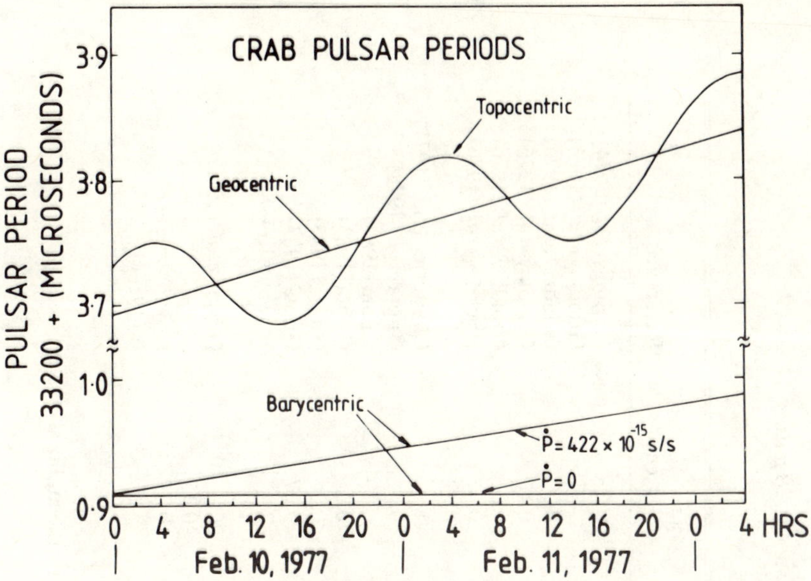

Figure 5.10. Variation of CRAB pulsar period with time. Periods at the solar system barycentric for d*P*/d*t* = 0 (hypothetical) and for d*P*/d*t* = 422 × 10^{-15} s s^{-1} (observed) are shown by the two lines at the bottom. Upper-most line: result of Doppler shift due to the Earth's orbital motion; sinusoidal line: result of Doppler shift due to the Earth's rotation. Note that the ordinate scale is vastly expanded.

5.2.3A CRAB Pulsar

The CRAB pulsar is known to emit in phase at radio, optical, X-ray and γ-ray frequencies; see Figure 4.11 and Manchester and Taylor (1977). The pulse profiles exhibit two peaks, corresponding to the main and the interpulses, separated by 0.42 in phase. Emission of UHEGR from this object has been studied in several experiments; see Table 5.2 for some representative results. In most cases there was no information on absolute phase. Where it existed, no constant relationship in phase was found to be present between the main radio pulse and the peak in UHEGR phase. The Ooty group (Gupta *et al*. 1978) observed two peaks in the phase plot of UHEGR with a mutual separation of 0.42 in phase, as was seen at lower energies. The UHEGR pulsed emission was not constant in time. The Mt. Hopkins and the Ooty groups, who observed this pulsar for several years, detected signals during some observations but not during others; see Table 5.2. The Iowa group (Erickson *et al*.

Table 5.2. *Results on pulsed emission of UHEGR from the CRAB pulsar*

Time-averaged fluxes.

Group	Date	Effect (σ)	$E_{\gamma\,min}$ (TeV)	Flux ($10^{-11}\,cm^{-2}\,s^{-1}$)	Reference
(A) Dublin	1970	3.5	2	0.2	Jennings *et al.* (1974)
(B) Mt. Hopkins	1969–72	—	0.12	<1.7	Grindlay *et al.* (1976)
	Feb. 1973	—	0.8	<0.2	Grindlay *et al.* (1976)
	Dec. 1973	5.2	0.8	0.4	Grindlay *et al.* (1976)
	Jan. 1974	4.3	1.8	1.3	Grindlay *et al.* (1976)
(C) Iowa	Mar. 1976	4.0	20	0.47*	Erickson *et al.* (1976)
				0.12	
(D) Ooty	1970	—	10	<1	Chatterjee *et al.* (1971)
	Feb. 1977	3.6, 2.2**	4.5	0.4* }	Gupta *et al.* (1978)
				0.29 }	Gupta (1983)
	1978	—	2.5	<0.2	Gupta (1983)
	1979	—	1.9	<0.2	Gupta (1983)
	1980	—	1.5	<0.27	Gupta (1983)
	Feb. 1981	4.2	0.9	2.4* }	Viswanath (1982)
				1.2 }	
(E) Durham	Oct. 1981	6.8	3	20* }	Gibson *et al.* (1982*b*)
				~0.15 }	
	1982–3	4.3	1.0	0.79	Dowthwaite *et al.* (1984*c*)

* When averaged over periods during which the effect was seen.

** Two significant peaks are reported in the phasogram by the authors.

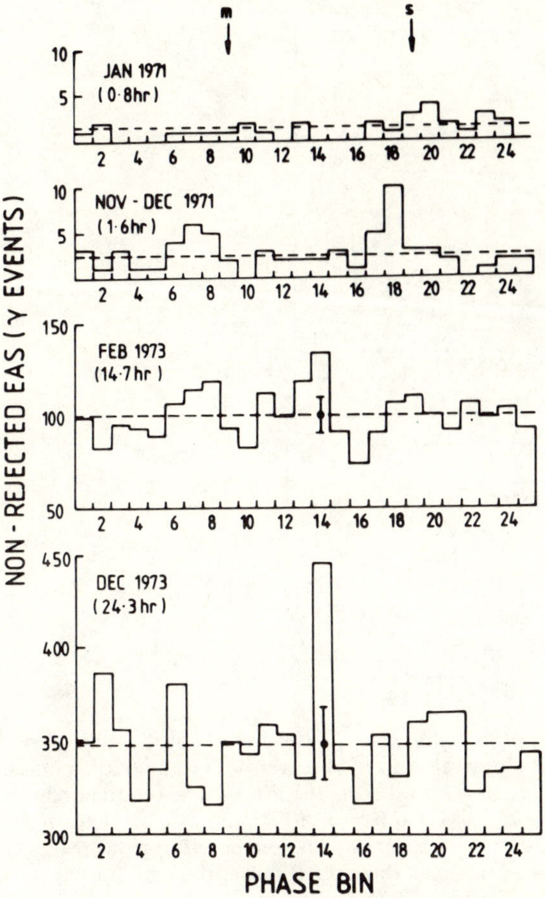

Figure 5.11. Phasograms (light curves) of the CRAB pulsar observed by Grindlay *et al.* (1976) at γ-ray energies greater than 800 GeV with the Mt. Hopkins 10 m reflector. Note the variability of amplitude and phase of the peaks in the distribution. The arrows indicate the radio main (m) and interpulse (s).

1976), the Durham group (Gibson *et al.* 1982*b*) and the Ooty group (Vishwanath 1982 and Vishwanath *et al.* 1985) find that the emission of UHEGR is variable over days, tens of minutes, minutes and even tens of milliseconds. The variability in UHEGR emission is illustrated in Figures 5.11 and 5.12 taken from Grindlay *et al.* (1976) and Gibson *et al.* (1982*b*), respectively.

The time-averaged energy spectrum of UHEGR from the CRAB pulsar has been presented by many authors; see, for example, Sreekantan (1981). Since then the Ooty group (Gupta 1983) – after detailed observa-

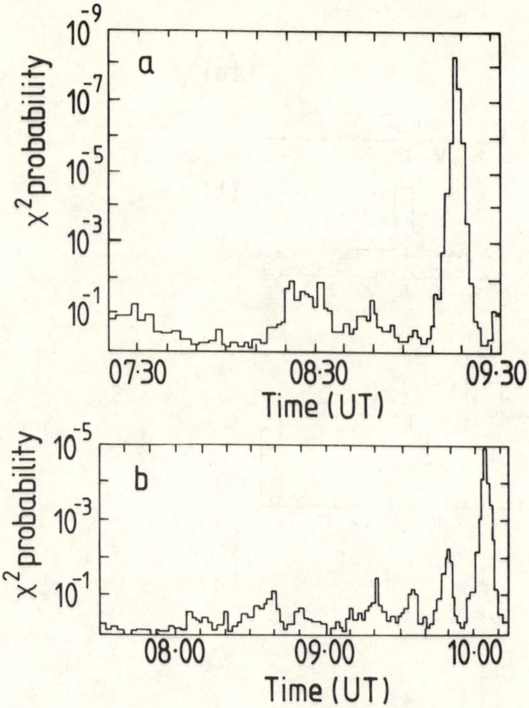

Figure 5.12. The probability of lack of periodicity at the radio period for data in 15 min intervals during observations on the CRAB pulsar by the Durham group (Gibson *et al.* 1982*b*). (*a*) For observations made on October 23, 1981. (*b*) For observations made on October 31, 1981. The peaks towards the right in both panels offer strong evidence that, on these two occasions, the CRAB pulsar did emit pulsed UHEGR over time spans ~15 min.

tions on the angular resolutions of their detector system – has revised the estimates of energy threshold and collection areas (both have increased). Table 5.2 lists their revised fluxes and threshold values and not the published values. We have shown, in Figure 5.13, the intensities or upper limits at various energies presented in Table 5.2. Also shown in the figure is the extrapolation of the energy spectrum observed at medium high energies by the COS B satellite (Lichti *et al.* 1980). From the figure two points emerge: (i) The intensities seem to show a scatter much larger than the uncertainties (a factor of ~2) in thresholds and collection areas. If this scatter is due to source variability, it is not clear if one can obtain and draw any inference from the energy spectrum. (ii) If one assumes that the true long term time-averaged intensities lie along the centroid of the scattered

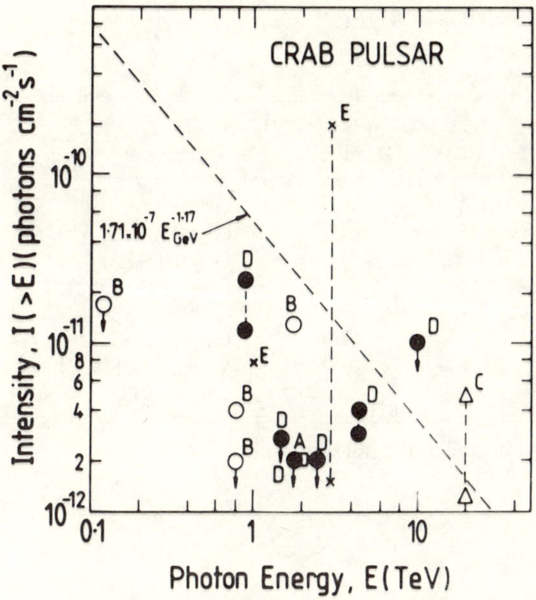

Figure 5.13. Upper limits and time-averaged fluxes from the CRAB pulsar as observed by the various groups: (A) Dublin; (B) Mt. Hopkins; (C) Iowa; (D) Ooty; and (E) Durham. See Table 5.2 for references. Upper values in the pairs of points connected by vertical lines refer to flux values when averaged only over periods during which the effect was seen and the lower values when averaged over the entire observation period. Slant dashed line: extrapolation of the COS B spectrum (Lichti *et al.* 1980).

points, they are clearly an order of magnitude lower than the extrapolated line. It appears that the spectral index in the integral power law spectrum steepens from -1.17 at $E_\gamma \sim 1$ GeV to ~ -1.5 at $E_\gamma > 1$ TeV. The time-averaged luminosity at $E_\gamma > 1$ TeV is $\sim 10^{34}$ erg s^{-1}, if one assumes that the emission is isotropic.

5.2.3B VELA pulsar

The VELA pulsar is located 500 pc away from the solar system. Being located in the southern skies it is accessible for observation only from sites near the equator or in southern latitudes and it has been observed so far only by two groups. Results on UHEGR emission are summarised in Table 5.3. The phasogram obtained by the Ooty group (Bhat *et al.* 1980) in their 1979 observations is shown in Figure 5.14. One sees two peaks, 4.4 and 2.2σ, separated by $\Delta\phi = 0.42$. It must be pointed

Table 5.3. *Results on pulsed emission of UHEGR from the VELA pulsar*

Time-averaged fluxes.

Group	Date	Effect (σ)	$E_{\gamma\,\mathrm{min}}$ (TeV)	Flux (10^{-11} $\mathrm{cm}^{-2}\mathrm{s}^{-1}$)	Reference
Narrabri	1972	4	0.3	1.0	Grindlay *et al.* (1975*b*)
	1973	3	5	0.07	Grindlay *et al.* (1975*b*)
Ooty	1977	—	40	<0.013	Gupta (1983)
	1979	4.4, 2.2*	14	0.05	Bhat *et al.* (1980) Gupta (1983)
	1980	—	10	<0.018	Gupta (1983)
	1981	4.2	7	0.07	Viswanath (1982) Gupta (1983)

* Two significant peaks are reported in the phasogram.

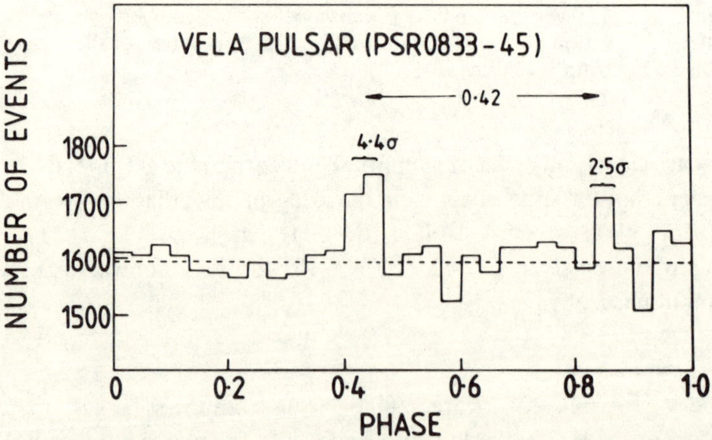

Figure 5.14. Phasogram of events observed from the direction of the VELA pulsar by the Ooty group (Bhat *et al.* (1980)). Dashed line: average background which is mostly due to cosmic ray-induced air showers. The two peaks separated by 0.42 in phase are of 4.4 and 2.5 σ in amplitude. The absolute phase is not known.

out, however, that at lower energies $\Delta\phi = 0.42$ was seen in the case of the VELA pulsar only in medium energy γ-rays (COS B), but not in the optical ($\Delta\phi = 0.24$) or radio (single peak), whereas, as already mentioned, $\Delta\phi = 0.42$ was seen at all frequencies and in phase in the case of the CRAB pulsar. The Ooty group has re-evaluated the collection areas and energy thresholds for the VELA search (Gupta 1983), and Table 5.3 lists the revised values of energy threshold and intensities. Data presented in Table 5.3 are shown in Figure 5.15. This pulsar too, like the CRAB, seems to be variable in the emission of UHEGR. The fluxes are generally three orders of magnitude lower (at $E_\gamma \sim 10$ TeV) than the extrapolated COS B spectrum (Kanbach *et al.* 1980). The spectral index of the integral energy spectrum, then, must steepen from -0.9 at $E_\gamma \sim 1$ GeV to -1.65 at $E_\gamma \sim 10$ TeV. The time-averaged luminosity at $E_\gamma > 10^{13}$ eV is $\sim 7 \times 10^{32}$ erg s^{-1}, if one assumes that the emission is isotropic.

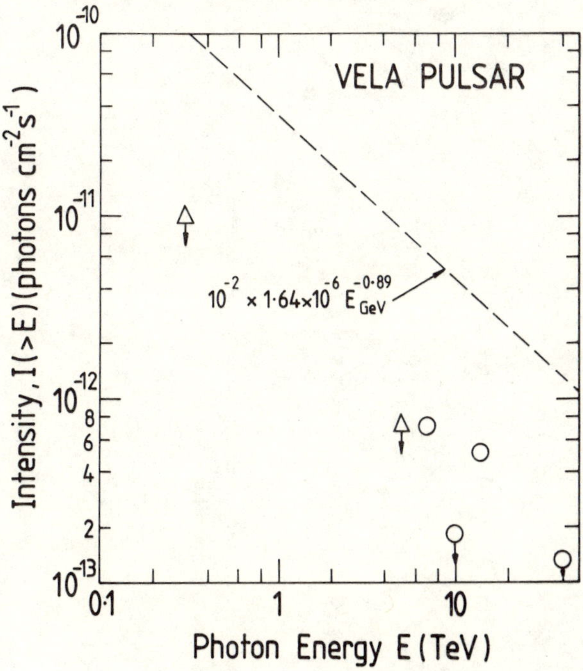

Figure 5.15. Upper limits and time-averaged fluxes from the VELA pulsar. Δ: from the Narrabri group (Grindlay *et al.* 1975*b*); O: from the Ooty group (Bhat *et al.* 1980, Vishwanath 1982, Gupta 1983). Slant dashed line: one-hundredth of the extrapolation of the COS B spectrum (Kanbach *et al.* 1980).

Table 5.4. *Results on pulsed emission of UHEGR from Cygnus X-3*

Time-averaged fluxes.

Group	Date	Effect (σ)	Phase	$E_{\gamma\,min}$ (TeV)	Flux ($10^{-11}\,cm^{-2}\,s^{-1}$)	Reference
Crimea	1972–7	5.4	0.15–0.20	2	18	Neshpor *et al.* (1979)
		3.0	0.78–0.83	2	9	
Mt. Hopkins	1976	—	—	0.3	<6.5	Weeks and Helmeken (1977)
Dublin/Mt. Hopkins	1980	3.5	0.6–0.8	2	15	Danaher *et al.* (1981)
Iowa/JPL/UC	1981	4.4	0.5–0.7	0.5	8	Lamb *et al.* (1982)
Durham	1981	4.1	0.625–0.655	1.3	20*	Dowthwaite *et al.* (1983)
	1982				0.3	

* Peak flux over 10 min interval.

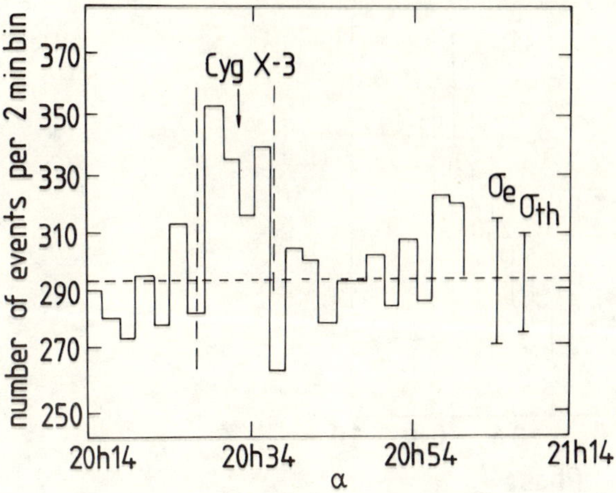

Figure 5.16. The first detection by the Crimean Astrophysical Observatory Group of UHEGR from Cygnus X-3: counts vs right ascension (Vladimirsky *et al.* 1973). σ_e: experimental standard deviation; σ_{th}: theoretical statistical error.

5.2.3C Cygnus X-3

Cygnus X-3, located at least 11.4 kpc from the solar system, is an X-ray binary system. It has been seen in medium energy γ-rays by two groups (Lamb *et al.* 1977, Galper *et al.* 1977) but not by COS B (Swanenburg *et al.* 1981). Results on UHEGR emission by this object are summarised in Table 5.4. The Crimean Astrophysical Observatory group (Vladimirsky *et al.* 1973, Neshpor *et al.* 1979, Stepanian *et al.* 1982) have carried out extensive observations on this source. Figure 5.16 (Vladimirsky *et al.* 1973) shows a drift scan in which it can be clearly seen that the object emits UHEGR. Later, the authors showed that the object pulsates with a period 0.199 683 d (~4.8 h) and also that the intensity is modulated with a 34.1 d period, also seen at X-rays. The 4.8 h period is interpreted as that of the binary orbital period while the 34.1 d period is thought to be due to precession. A phasogram of UHEGR is shown in Figure 5.17 (Danaher *et al.* 1981). An examination of Table 5.4 reveals that (i) the intensity is variable in time, and (ii) while the Crimean group sees emission in two peaks at phases ~0.18 and 0.80 (X-ray minimum is taken as zero phase), the other groups see it in only one peak at ~0.5 to 0.8. Even among the latter, there are finer differences in the phase which perhaps arise from using differing X-ray ephemerides (Porter 1983).

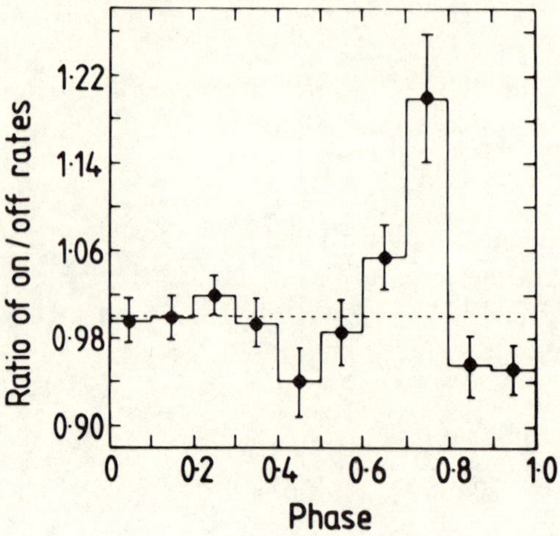

Figure 5.17. Phase histogram of UHEGR emission for Cygnus X-3 from the work of Danaher *et al.* (1981).

Neshpor *et al.* (1979) deduce that the luminosity of the source at energies greater than 2 TeV is 1.2×10^{37} erg s^{-1} during the first peak ($\phi \sim 0.18$) and 6×10^{36} erg s^{-1} during the second ($\phi \sim 0.80$), assuming isotropic emission.

5.2.3D Hercules X-1

This object is a pulsating and eclipsing X-ray binary located at a distance of 5 kpc from the solar system. In optical light, and in X-rays, the source shows modulations on time scales of 1.24 s, 1.7 d and 35 d. The Durham group, Dowthwaite *et al.* (1984*d*), observed this source on April 17, and during July 3–13, and October 7–11, 1983, During one of the four scans on April 17, the authors saw a $>3\sigma$ count rate excess for 3 min periodic at a $>3\sigma$ level at the X-ray period. The peak γ-ray flux at $E_\gamma > 1000$ GeV during the burst activity was 1.2×10^{-9} cm^{-2} s^{-1}. The outburst occurred at a phase of 0.76 in the 1.7 d period and 35 d before a detected turn-on in X-rays.

The July data, however, showed an overall periodicity only at the 2σ level. The best period determined by the authors from their own data, $P = 1.237\,787$ s, agrees well with the one derived from the contemporary X-ray measurements. The time-averaged flux is given by the authors as $(3 \pm 1.5) \times 10^{-11}$ cm^{-2} s^{-1} at $E_\gamma > 1$ TeV.

The time-averaged luminosity at $E_\gamma > 1$ TeV from Her X-1 is computed by the authors to be 2×10^{35} erg s^{-1}, a value which is 0.5 % of the X-ray luminosity, including soft X-rays.

5.2.3E Other pulsars

Zyskin and Mukanov (1983) observed 2CG195 + 4 (Geminga), a γ-ray source seen strongly at medium energies by both SAS 2 and COS B (see Chapter 4). Though the source was not seen at radio frequencies, it was seen in X-rays (Caraveo *et al.* 1984, Bignami *et al.* 1984). The authors find that UHEGR at $E_\gamma > 10^{12}$ eV during 1981, show a periodicity of 59.28 $\pm$ 0.01 s, a value in reasonable agreement with that reported at medium energies. The authors estimate the integral flux of UHEGR from this object to be $(5 \pm 3) \times 10^{-11}$ cm^{-2} s^{-1} at $E_\gamma > 10^{12}$ eV. In an earlier observation, Helmken and Weekes (1979), not seeing a signal, placed upper limits of 1.6×10^{-10} and 6×10^{-11} cm^{-2} s^{-1} at $E_\gamma > 0.2$ and 1 TeV, respectively. The result of Zyskin and Mukanov is statistically marginal and needs to be confirmed in an independent experiment.

Chantler *et al.* (1984) observed PSR 1937 + 21, a 'millisecond pulsar' with $P \sim 1.558$ ms, and found that the light curve at $E_\gamma > 10^{12}$ eV showed a single wide ($\Delta\phi \sim 0.4$) peak. The authors find that the best period determined by them is lower by 6 parts in 10^7 when compared with the radio period. This discrepancy is outside the errors deduced at radio and UHEGR frequencies. The origin of the discrepancy is not clear. The authors deduce a peak flux of 3×10^{-10} cm^{-2} s^{-1} at $E_\gamma > 10^{12}$ eV. The signal was not seen on all occasions when the source was observed. This result is yet to be confirmed by an independent experiment.

Cawley *et al.* (1983) observed PSR 1953 + 29 ($P \sim 6.1$ ms); the data await analysis.

We now turn our attention to the possible emission of UHEGR by primordial black holes – a hypothetical class of celestial objects whose existence is yet to be confirmed.

5.2.3F Primordial black holes

According to Page and Hawking (1976), it is conceivable that the hot and highly compressed regions, believed to exist shortly after the 'big bang', might have collapsed into black holes. These objects are called primordial black holes (PBH) as distinct from stellar black holes. PBH in the mass range 10^{14}–10^{15} g would have survived until the present epoch. Their temperatures are as high as 10^{12} K and the event horizon radius is as small as 1 Fermi or even less. With such small confines, quantum mechanical effects come into play and lead to particle emission by PBH through spontaneous creation of pairs of particles near the event horizon. As the particle emission progresses, the mass of the PBH decreases, raising its temperature and leading to even more copious production of particles. The system is thus clearly in a run-away state; for details, see

Page and Hawking (1976) and Carr (1976). The PBH with masses $\sim 5 \times 10^{14}$ g, should they exist, must be evaporating at this epoch so fast as to effectively explode. According to the elementary particle mode of evaporation, $\sim 10^{30}$ γ-rays at $E_\gamma > 10^{12}$ eV are emitted over a period of 0.1 s.

Porter and Weekes (1979) looked for seven or more Cerenkov flashes ($E_\gamma > 3 \times 10^{12}$ eV) occurring in less than 0.1 s. Not finding any, the authors set an upper limit (99 % confidence level) of 3×10^4 pc^{-3} y^{-1} to PBH explosions. In experiments at mountain altitudes Bhat *et al.* (1982) looked for bursts of low energy photon showers ($\geqslant$three showers within a time span of a few tens to a few hundreds of milliseconds) produced by 5 TeV γ-rays but did not find any. The authors set a 99 % confidence level upper limit to PBH explosions of 1.6×10^3 pc^{-3} y^{-1}.

5.3 Experiments at $E_\gamma > 100$ TeV

5.3.1 The air shower technique

At energies above 10^{14} eV, the electromagnetic cascade showers initiated in the terrestrial atmosphere by celestial γ-rays can penetrate down to sea-level, though it is ~28 radiation lengths deep. At $E_\gamma > 10^{14}$ eV, the densities of $e^\pm$ in the shower at sea-level are small (10–0.1 m^{-2}) at distances 1–100 m from the core of the shower. It is, therefore, more convenient to work with γ-rays at $E_\gamma > 10^{15}$ eV, which result in higher densities of $e^\pm$. One samples at sea-level or mountain altitudes particle densities at several points on a horizontal plane by using detectors (usually of dimensions ~1 m^2) spread out to distances of several hundred metres. The technique was primarily developed for studying the charged component of primary cosmic rays at energies $E_0 > 10^{14}$ eV. We denote these latter showers by EAS (extensive air showers), while referring to the former as γ-ray showers. In EAS one has, besides $e^\pm$, muons (~5 % of $e^\pm$), pions, nucleons and other hadrons (~1 % of $e^\pm$). In EAS experiments detectors are also employed to study $\mu^\pm$ and hadrons – their identity, energy, number and lateral distribution – in order to gain an insight into high energy interaction characteristics and the cosmic ray primary composition. This technique, which is four decades old, has been described by several authors; see, for example, Galbraith (1958) and Greisen (1960). Unlike the experiments based on the atmospheric Cerenkov technique, the experiments with particle detector arrays can be carried out day and night and in any weather, making the duty factor much higher.

To estimate the shower size, N, one fits to the sampled particle densities a lateral distribution function of the type given by Nishimura and Kamata (1952):

$$\rho(r) = \frac{N}{2\pi r_1^2} f(s)\left(\frac{r}{r_1}\right)^{s-2}\left(1 + \frac{r}{r_1}\right)^{s-4.5} \tag{5.4}$$

Here $\rho(r)$ is the density of particles m^{-2} at a distance r from the shower core; s is the age parameter of the shower; $f(s)$ is a function of s; and r_1 is the Moliere scattering unit (~80 m at sea-level). From such a fit the core location is determined, together with the age parameter and shower size, the latter being given by

$$N = \int_0^\infty 2\pi r \rho(r)\,\mathrm{d}r \tag{5.5}$$

The energy, E, of the primary charged cosmic ray or the γ-ray is found from N by adopting the E–N relationship computed from the known (or extrapolated where necessary) cross sections for various strong and electromagnetic interaction processes.

The arrival direction of the primary cosmic/γ-ray is obtained from fast timing information obtained by suitably designed detectors. If, for example, one employs two timing detectors separated by a distance d on a horizontal plane, the projected zenith angle, θ_p, in a vertical plane containing the two detectors is given by

$$\theta_\mathrm{p} = \sin^{-1}(tc/d) \tag{5.6}$$

Here c is the velocity of light and t is the difference in arrival times of the shower front at the two detectors. For typical values of $\theta_\mathrm{p} \sim 30°$, $d \sim 50$ m and Δt (error in timing measurements) ~ 3 ns, one can measure θ_p to an accuracy $\Delta\theta_\mathrm{p} \sim 1°$. Using several non-collinear timing detectors for redundancy and improved directional accuracy, it is possible to determine the spatial zenith angle and azimuthal angle of the arrival direction of the shower. This information, together with that on event time, enables the shower direction in celestial coordinates to be found. If there is enough redundancy built into the timing system, it is also possible to solve for the finite radius of curvature ($R \sim 3$ to 5 km) of the shower front, instead of assuming it to be flat.

The EAS (which are the background to the UHEGR signal) are overwhelmingly large in number compared to γ-ray showers, and it is imperative to first reduce the EAS background and then to look for spikes either in the spatial or temporal distributions of the recorded events. Such spikes can, then, be attributed to γ-rays.

To reduce the EAS background, use is made of two entities: (i) the age parameter, s, of the shower, and (ii) the muon content. The energy in an electromagnetic cascade, as it penetrates the terrestrial atmosphere, is attenuated much faster than that in a nuclear–electromagnetic cascade in which the energy is carried forward through the relatively stable hadrons and the follow-through nucleons. As a result, γ-ray showers look much older ($s > 1$), i.e. the lateral distribution of the electrons and positrons in the shower is much flatter than in the EAS ($s \sim 1$). Thus, a cut is made on s, e.g. $s > 1.1$, to enrich the data sample in UHEGR.

Next to $e^{\pm}$, the most abundant charged particles in an EAS are $\mu^{\pm}$. In γ-ray showers there are far fewer muons than in EAS. The production cross section for a muon pair in the Coulomb field of nuclei by a γ-ray is much lower than that of an $e^{\pm}$ pair, being given by

$$\sigma(\mu^{\pm})/\sigma(e^{\pm}) = (m_e/m_{\mu})^2 = 2.3 \times 10^{-5} \tag{5.7}$$

More relevant for muon production in γ-ray showers is the photoproduction of charged pions which subsequently decay into $\mu^{\pm}$. Even here, in collisions with 'air' nuclei, the corresponding ratio

$$\sigma(\gamma + N \rightarrow \pi^{\pm} + \ldots)/\sigma(p + N \rightarrow \pi^{\pm} + \ldots) \approx 10^{-2} \tag{5.8}$$

is very small and one expects, therefore, far fewer $\mu^{\pm}$ in a γ-ray shower than in an EAS. Wdowczyk (1965) and Braun and Sitte (1965) have made detailed calculations and have shown that the ratio of the number of muons in a γ-ray shower to that in a conventional EAS is ~0.1. One, then, looks for an anomalously low muon density as an additional signature of a γ-ray shower.

5.3.2 Results

Very early in the searches for celestial γ-rays at $E_{\gamma} > 10^{14}$ eV, Gawin *et al.* (1963) and Toyoda *et al.* (1965) found that a small fraction of showers contained anomalously low numbers of muons (mu-poor showers). However, it was not clear if the 'mu-poor' criterion alone was sufficient to qualify a shower as due to a γ-ray primary because there could be fluctuations in the characteristics of high energy p–N collisions, in effect causing almost all the energy to be transferred to the π^0 component. If this happened, the shower, though created by a charged cosmic ray primary, would appear as one created by a γ-ray. Also, one could not pin-point any source direction with which such an excess was associated.

These early experiments either did not have a timing facility or, if they did, the angular resolution was not good enough. The deficiency was remedied at least in some of the later experiments.

5.3.2A Cygnus X-3

Recently, the Kiel group (Samorski and Stamm 1983*a*, *b*, Stamm and Samorski 1983) have published impressive evidence for having detected pulsed UHEGR at $E_\gamma > 10^{15}$ eV from Cygnus X-3. These authors operated a 28 scintillator array (each 1 m^2 in area) at Kiel for four years (1976–9). They also recorded fast (ns) timing information from 22 detectors, which enabled them to determine the shower direction to an accuracy of ~1°. The authors could determine for each shower the size, N (to an accuracy of ±30 %), core location (±1 m) and age parameter, s (±0.1).

Stamm and Samorski plotted the shower directions in bins of ~4° × 3° in a right ascension–declination grid. Although they did not find an excess in the Cygnus X-3 direction compared to the background among the showers with $s < 1.1$, they found a 4.4σ excess for (γ-ray enriched) showers with $s > 1.1$; specifically they found 31 showers when the average for the bin was 14.4. The probability that the average fluctuated to 31 or more is $\sim 10^{-4}$. More significantly, Samorski and Stamm (1983*b*) found that the event times exhibited a periodicity with peak emission at a phase of 0.35. The authors scanned the period in the interval from $(P - 5 \times 10^{-5})$ d to $(P + 5 \times 10^{-5})$ d in steps of 10^{-7} d. Here $P = 0.199\,681\,4$ d at an epoch t_0 = JD 2 440 949.917 6 (and $\dot{P} \equiv 0$) given by Parsignault *et al.* (1976), based on their X-ray data. The result is shown in Figure 5.18. It is significant that the best value determined from these data, $P = (0.199\,681\,6 \pm 2 \times 10^{-7})$ d, agrees within the errors with that found at X-ray energies. The authors calculated the light curves (phasograms) for the 31 showers using their best value of P for 10, 50, 100 and 150 bins. The result is shown in Figure 5.19. It is remarkable that the pulsed emission is concentrated over a very narrow phase interval, $\Delta\phi = 7 \times 10^{-3}$. The authors give a time-averaged flux of $(7.4 \pm 3.2) \times 10^{-14}$ cm^{-2} s^{-1} at $E_\gamma > 2 \times 10^{15}$ eV.

Turning to the question of luminosity, Cawley and Weekes (1984), assuming a distance to the source of 11.4 kpc, correct the above flux for absorption due to the interaction of UHEGR with the 2.7 K black body radiation (a correction factor of 3.3) and estimate the luminosity at $E_\gamma > 2 \times 10^{15}$ eV to be 2×10^{37} erg s^{-1}. This value may be compared with luminosities of this object at X-rays and at all energies $>10^5$ eV of 10^{37} and

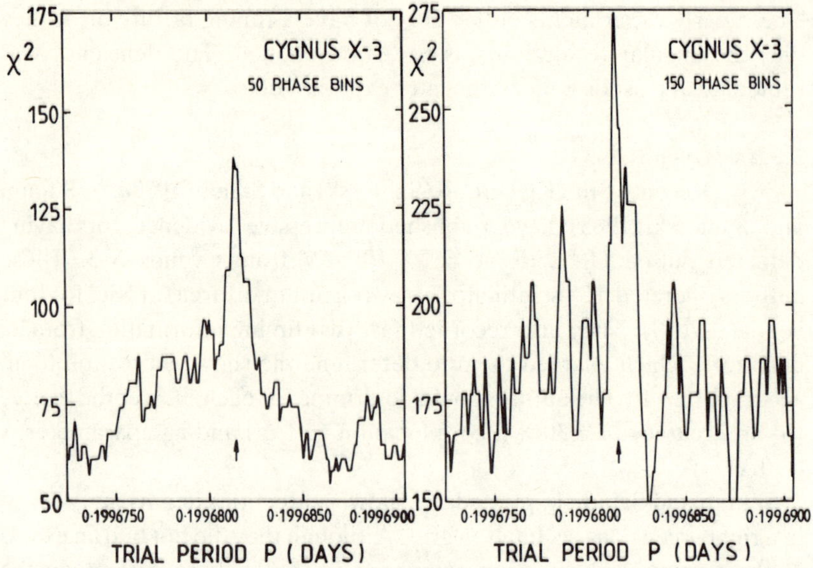

Figure 5.18. χ^2 tests to the phase of Cygnus X-3, in the data observed by Samorski and Stamm (1983*b*), as a function of trial period P for subdivisions of the phase into 50 (left panel) and 150 (right panel) bins. Arrows: indicate period found by Parsignault *et al*. (1976) from X-ray data. The highest χ^2 (for the null hypothesis) determines the best value for the period, which is in close agreement with the one determined from X-ray data.

2×10^{38} erg s^{-1}, respectively. Samorski and Stamm (1983*a*) deduced the energy spectrum of γ-rays to be

$$N(>E_\gamma) = (2.5 \pm 0.9) \times 10^{-7} E_\gamma(\mathrm{GeV})^{-1.128\pm0.013}\ \mathrm{cm}^{-2}\ \mathrm{s}^{-1} \quad (5.9)$$

by combining all the data in the range 10^8–10^{16} eV. The authors should have seen several events above 2×10^{16} eV on the basis of extrapolation of the above spectrum, but they saw none; thus, there is probably a cut-off or steepening of the spectrum above about 2×10^{16} eV.

Samorski and Stamm (1983*c*) have also observed muons in the showers, the ratio of average muon densities at 10 m distance from the shower core being given as

$$\langle \rho_\mu(10\ \mathrm{m})_{\mathrm{on}} \rangle / \langle \rho_\mu(10\ \mathrm{m})_{\mathrm{off}} \rangle = 0.77 \pm 0.09 \quad (5.10)$$

Here the subscript 'on' refers to showers from the source direction (γ-rays with negligible EAS background), and 'off' to those from other directions

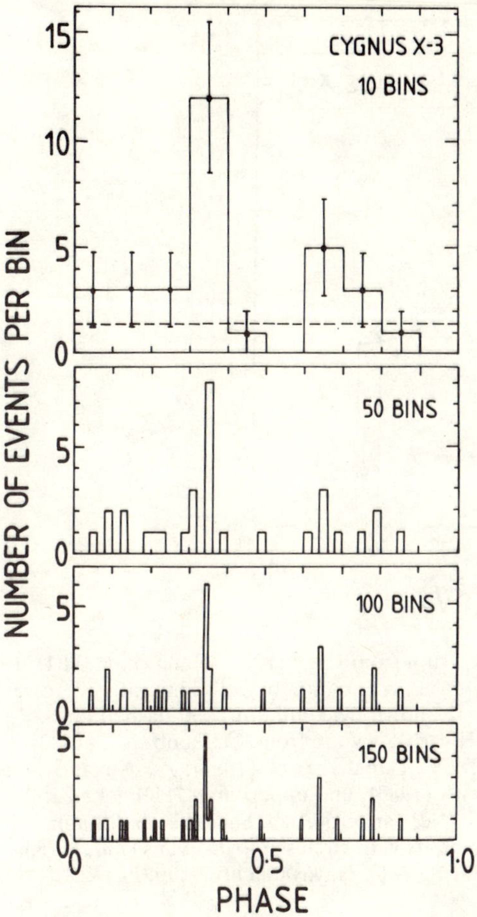

Figure 5.19. The Cygnus X-3 data of Samorski and Stamm (1983*b*) subdivided into different numbers of bins. Dashed line: average off-source rate of 1.44 ± 0.04 showers per 0.1 of phase.

(all are EAS). This ratio is much larger than the value (~0.1) expected for γ-ray initiated showers – a fact hard to understand.

Lloyd-Evans *et al.* (1983*a, b*) observed Cygnus X-3 for the four years 1979–82 with the Haverah Park (UK) air shower array. These authors too find an excess of showers (1.7σ) in the direction of Cygnus X-3. When they analyse the event times of showers in the Cygnus X-3 direction for periodicity using the ephemeris of van der Kliss and Bonnet-Bidaud (1981), they find that the phasogram shows a peak in the phase interval 0.225 to 0.250. They see 73 events in this phase bin, whereas the average

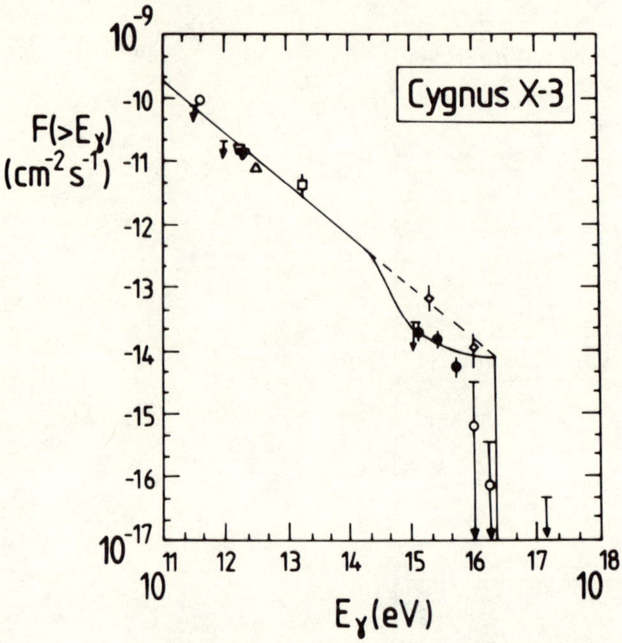

Figure 5.20. γ-ray spectrum from Cygnus X-3 (Rana *et al.* 1984). Best fit spectra with and without attenuation by collisions with microwave background photons are shown by continuous and dashed lines, respectively. Data at $E_\gamma < 10^{14}$ eV are from: ○, Lamb *et al.* (1982); ▽, Stepanian *et al.* (1982); ▼, Danaher *et al.* (1981); △, Mukanov *et al.* (1980); □, Morello *et al.* (1983), and upper limits, Helmeken *et al.* (1979). Data at $E_\gamma > 10^{14}$ eV are from: ◇, Samorski and Stamm (1983*a*); ● and upper limits with circles, Lloyd-Evans *et al.* (1983*a*); upper limits at 10^{15} and 10^{17} eV, Hayashida *et al.* (1981).

is 39. The statistical chance of observing such a peak in any one of the 40 bins is 2.8×10^{-5} and anywhere in the entire plot, $40 \times 2.8 \times 10^{-5} \sim 10^{-3}$. The authors give an integral flux of $(1.5 \pm 0.3) \times 10^{-14}$ cm^{-2} s^{-1} at $E_\gamma > 3 \times 10^{15}$ eV, which is a factor ~3 lower than the Stamm and Samorski flux values; see Figures 5.20 for a composite spectrum of UHEGR from Cygnus X-3, taken from Rana *et al.* (1984). By examining all the available data at the various wavelengths Rana *et al.* (1984) conclude that the output of Cygnus X-3 is probably decreasing with time.

Lloyd-Evans *et al.* find, in agreement with the Kiel group, that the spectrum steepens at $E_\gamma > 3 \times 10^{16}$ eV. These authors have presented an interesting circular graph, shown in Figure 5.21, in which are presented the phases and statistical probabilities for chance fluctuations of back-

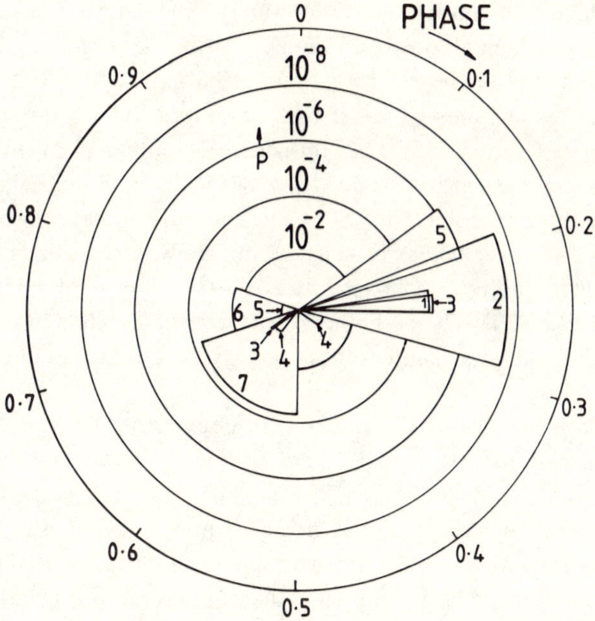

Figure 5.21. A summary of phases (relative to X-ray minimum) for pulsed emission above 500 GeV, as given by Lloyd-Evans *et al.* (1983*a*). The azimuth represents phase, and circles the magnitudes of probabilities that the peaks are due to chance fluctuations. Numbers 1–7 refer to the works of Lloyd-Evans *et al.* (1983*a*), Samorski and Stamm (1983*a*), Samorski and Stamm (1983*b*), Morello *et al.* (1983), Stepanian *et al.* (1982), Danaher *et al.* (1981), and Lamb *et al.* (1982). Note that Lloyd-Evans *et al.* have subtracted 0.11 from the phase values published by Samorski and Stamm (Sections 2 and 3 in this figure) so as to comply with the ephemeris used by the former.

ground to account for the signal in all the cases claimed by the various authors so far. Lloyd-Evans *et al.* did not present any data on muon densities.

Morello *et al.* (1983), using their small air shower set-up at Plateau Rosa (3500 m above sea-level), find rather weak evidence for UHEGR emission at the modest 2.8σ level. The phasogram shows a peak in the phase range 0.60–0.65. The estimated flux is $(4.2 \pm 1.5) \times 10^{-12}\ \mathrm{cm}^{-2}\,\mathrm{s}^{-1}$ at $E_\gamma > 3 \times 10^{13}$ eV.

Bhat *et al.* (1985) have analysed data taken during 1976–7 in an experiment that detected atmospheric Cerenkov light produced by extensive air showers and UHEGR at energies greater than 5×10^{14} eV. The experiment was carried out at Gulmarg, India, at an altitude of 2743 m above sea-level. It consisted of recording coincidences between two photomultipliers facing upward to the sky during clear moonless nights. The field of view was large, with a cone of half-angle 70°. The authors analysed the event times to see if they exhibited the 4.8 h periodicity appropriate to Cygnus X-3. The 'on-source' events, i.e. events in the approximate right ascension range, 1600 to 0000 h, exhibited a 4.5σ peak in the phase interval $\phi = 0.55$–0.60, while the 'off-source' events (in the right ascension range 0000 to 1600 h) did not show a peak. From this the authors deduced a flux of $(1.6 \pm 0.4) \times 10^{-12}$ cm^{-2} s^{-1} for UHEGR at $E_\gamma > 5 \times 10^{14}$ eV. This flux is higher, by a factor of 4.7, than those shown in Figure 5.20, and may again point to a decreasing intensity of emission with time.

Hayashida *et al.* (1981) did not detect γ-rays from Cygnus X-3 and quoted an upper limit of $\sim 3 \times 10^{-14}$ cm^{-2} s^{-1} at $E_\gamma > 10^{15}$ eV.

To sum up, the evidence of UHEGR signals presented by the Kiel and Haverah Park groups is quite convincing, although one does not understand at this stage why the muon densities in such γ-ray initiated showers in the Kiel data are so large and why Hayashida *et al.* (1981) did not see the signal. Another point of concern is that different authors used different X-ray ephemerides in their data analysis. The peaks in the phasograms are strong only if a particular X-ray ephemerides is used in the data analysis. Thus, although the Kiel and Haverah Park measurements give some support, one to the other, this support is not as strong as might have been hoped for.

5.3.2B CRAB nebula/pulsar

Dzikowski *et al.* (1983) observed this source during 1968–71 and again, during 1975–82, using an air shower array. There was no fast timing information and, hence, the shower arrival directions could not be precisely determined. Since the air shower intensity drops with increasing zenith angle, the event time itself defines a broad right ascension region from which a shower arrived. When the authors sorted out their data into bins, of width 3 h and 2.5 h RA, they found an excess of 58.8 ± 14.6 showers riding a background of 115.2 events in the RA bin centred on the CRAB nebula/pulsar region (5 h 30 min RA). This excess amounted to a flux of 2×10^{-13} cm^{-2} s^{-1} at $E_\gamma > 2 \times 10^{16}$ eV. Since the phase analysis of the event times was not given, one does not know if the excess showers are due to a pulsed component from the pulsar or due to steady emission from nebula/pulsar. It may be pointed out that, since the source direction is known only crudely, $\Delta\theta \sim 30°$, in this experiment one cannot be sure that the UHEGR are coming from the CRAB. When this flux is combined with the fluxes around 10^{12} eV from atmospheric Cerenkov work, the spectral index of the differential energy spectrum has a value -1.4, i.e. the spectrum is very flat. Dzikowski *et al.* find that the muon densities in the showers from this bin are 60 % of those from the other bins, a high value for primary γ-rays, as in the Kiel experiment. The luminosity of the source at $E_\gamma > 2 \times 10^{16}$ eV is estimated to be $\sim 3 \times 10^{36}$ erg s^{-1}. The authors add that although the γ-ray origin of the showers is preferred it is just possible that the primaries are instead protons coming from the early stage of a supernova explosion; however, a remarkably low Galactic magnetic field is required for such a hypothesis.

Boone *et al.* (1984) observed the CRAB nebula/pulsar region on December 9, 1980, and again on three nights in February, 1981, using their 'Fly's Eye' device to detect Cerenkov flashes initiated by cosmic ray/γ-ray primaries. The data were binned in 1 h wide right ascension bins based on event times (individual shower arrival directions could not be determined). From the data taken on December 9, 1980, the authors found 34 ± 11 excess showers (a 3.1σ effect) in the RA bin centred on the CRAB nebula/pulsar. This yielded a flux of $(2.1 \pm 0.7) \times 10^{-12}$ cm^{-2} s^{-1} at $E_\gamma > 10^{15}$ eV. No signal was found in the February 1981 data, corresponding to an upper limit of 5.3×10^{-13} cm^{-2} s^{-1} on the flux at $E_\gamma > 10^{15}$ eV, i.e. a factor 4 lower than the flux seen in the same experiment two months earlier. There is clearly the suspicion again that UHEGR emission from the CRAB nebula/pulsar is variable. The integral energy spectrum from the CRAB, as summarised by Boone *et al.* (1984), is shown in Figure 5.22. Other workers have also found only upper limits to the CRAB flux; Hayashida *et al.* (1981) quote an upper limit of 3×10^{-13} cm^{-2} s^{-1} at $E_\gamma > 10^{15}$ eV, and Craig *et al.* (1983) place an upper limit of 10^{-13} cm^{-2} s^{-1} at $E_\gamma > 3 \times 10^{15}$ eV.

On balance, evidence for the emission of UHEGR from the CRAB nebula/pulsar region at $E_\gamma > 10^{15}$ eV appears less convincing than that from Cygnus X-3.

5.3.2C VELA X-1

VELA X-1 is an X-ray binary located 1.4 kpc away from the solar system. The orbital period is 8.964 d. The object has been seen, as its name implies, in X-rays but not as yet in medium energy γ-rays. Protheroe *et al.* (1984) observed VELA X-1 for three years (1979–81) using the Buckland Park air shower array in Australia. With this array the arrival directions of showers were known from fast timing, and showers were

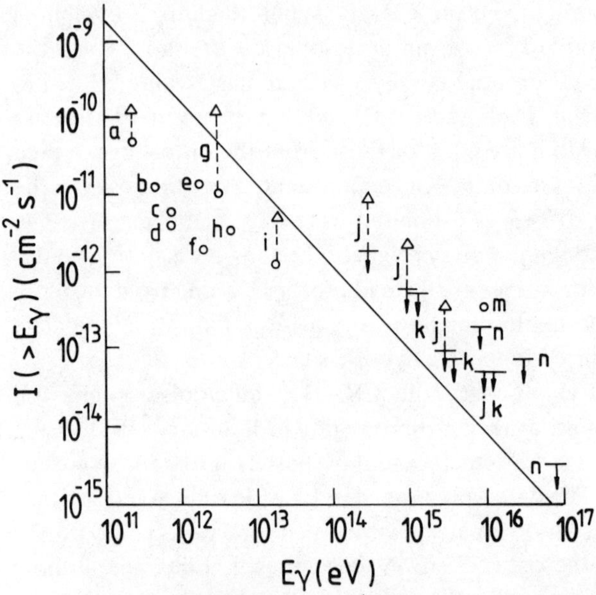

Figure 5.22. Spectrum of UHEGR from the CRAB pulsar region (Boone *et al.* 1984). △: 'high flux' observations made during single nights; ○: measurements averaged over longer times. Above 10^{14} eV, upper limits are also shown. Where experiments obtained both 'high flux' and longer term averages or upper limits, dashed lines connect the pairs of results. Solid line: extrapolation of the spectrum of McBreen *et al.* (1973), namely $I\,(>E_\gamma) = 0.95\ E_\gamma^{-1.08}$ keV for average fluxes of 1.5 keV to 2 GeV γ-rays. (*a*) Fazio *et al.* (1972); (*b*) Gupta *et al.* (1978); (*c–e*) Grindlay *et al.* (1976); (*f*) Jennings *et al.* (1974); (*g*) Gibson *et al.* (1982); (*h*) Porter *et al.* (1976); (*i*) Erickson *et al.* (1976); (*j*) Boone *et al.* (1984); (*k*) Hayashida *et al.* (1981); (*m*) Dzikowski *et al.* (1981); and (*n*) Craig *et al.* (1981).

pre-selected by making a cut on the age parameter, $s > 1.3$, to enrich the data sample in γ-rays. When the authors attempted a periodicity test using 10 bins with the period known from X-ray data, the signal was marginal but, when they binned the data into 50 bins, the authors found an apparent signal: eight showers were seen at the phase $\phi = 0.65$ when the expectation was 1.5 events. The authors concluded that at the 99.7 % confidence level the events were not due to fluctuations in the EAS background.

The time-averaged flux of $(9.3 \pm 3.4) \times 10^{-15}\ cm^{-2}\ s^{-1}$ is quoted, and this corresponds to a luminosity $\sim 2.3 \times 10^{34}\ erg\ s^{-1}$ (at $E_\gamma > 3 \times 10^{15}$ eV), which is about 1 % of that from Cygnus X-3. More observations on VELA X-1 are very desirable.

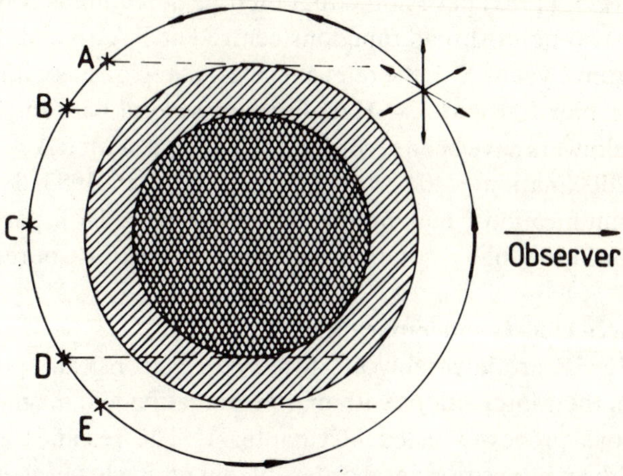

Figure 5.23. Schematic representation of how the UHEGR light curve can be generated from the Cygnus X-3 binary system (Vestrand and Eichler 1982). The cross-hatched region denotes the main body of the companion star, and the circumscribed shaded region, its atmosphere. The outer circle represents the pulsar orbit and the dashed lines, particle trajectories that produce γ-rays detectable at Earth. Pulses are produced when the pulsar is between A and B and between D and E. X-ray minimum occurs at position C.

5.3.2D Other sources

The Kiel group (Stamm and Samorski 1983) have reported evidence for the emission of UHEGR at $E_\gamma > 10^{15}$ eV from five other sources (besides Cygnus X-3) at the 3σ level. The fluxes are generally $\sim 2 \times 10^{-13}$ cm^{-2} s^{-1} at $E_\gamma > 10^{15}$ eV. The authors state that the observations are statistically not as significant as those on Cygnus X-3 and they await confirmation in a subsequent experiment.

Stamm and Samorski have also looked at six of the COS B sources, but without success.

The Gulmarg experiment has given rise to further tantalising results very recently. In addition to the pulsed Cygnus X-3 emission, Bhat *et al.* (1985) also found that the phase-averaged event rate is ~20 % higher for their 'on-source' histogram (RA = 14–02 h), which is consistent with their earlier result (Bhat *et al.* 1980), where they found that the hourly rates of the Gulmarg Cerenkov events displayed a broad peak (amplitude ~20 %) in the range 20 ± 04 h. Though the poor angular resolution of the Gulmarg experiment prevents seeking any definitive associations at this stage, Bhat *et al.* (1985) have noted that their peak feature is compatible with *at least* two point-spread functions centred at RA 16 and 20 h and that, apart from Cygnus X-3, the relevant celestial region has within it two Kiel sources, Her X-1 and AM-Her. Further, several hadron-poor and muon-poor showers have been recorded at Tien-Shan with RA ~ 16 h and ~18 h and declination ~40° ± 10° (Stamenov *et al.* 1983, Nikolsky, private communication), making the RA range 20 ± 04 h particularly promising for future high resolution searches for UHEGR sources.

5.4 Source models and implications

UHEGR are invariably produced by electrons (and positrons) and nuclei in their interactions with matter, radiation and magnetic field by the various processes listed in Chapter 1. The result is a flux of UHEGR with an intensity several orders of magnitude lower than that of the primary cosmic rays. As was mentioned in Section 5.1, it has not yet been possible to measure such a low intensity of the diffuse UHEGR in the presence of the very considerable cosmic ray background. At this stage, therefore, we can consider only point sources of UHEGR.

The point sources are considered in Sections 5.4.1 and 5.4.2 and the discernible implications in the areas of cosmic radiation and X-ray astronomy are touched upon in Section 5.4.3.

5.4.1 Production of UHEGR in supernova remnants

As was mentioned in Section 5.1, there exists indirect evidence from studies of the optical continuum from the CRAB nebula that high energy $e^{\pm}$ are present in the envelope of this SNR. Their expected interactions can be considered following the analysis of Ginzburg and Syrovatskii (1964). The characteristic synchrotron photon energy, E_γ, emitted by an electron of energy, E_e, in a transverse magnetic field, $H_\perp$, is given by

$$E_\gamma(\text{eV}) = 5 \times 10^{-9} H_\perp(\text{G})(E_e/m_e c^2)^2 \qquad (5.11)$$

Thus, an electron of 10^{14} eV moving in a transverse magnetic field of 3×10^{-4} G, such as experienced in the remnant, would produce 60 keV X-rays.

The lifetime of electrons against synchrotron emission, given by

$$t_{1/2} = 5 \times 10^8 (m_e c^2/E_e) H_\perp^{-2} \qquad (5.12)$$

is quite small, however, and in the example cited above $t_{1/2}$ is only about one year. It is clear, therefore, that high energy electrons in the CRAB nebula cannot have survived since the time of the initial explosion, an event which occurred 930 years ago. Instead, the electrons must be accelerated on a continual basis even at the present time. If the electrons in the SNR are secondary to protons, in the sense that they are produced through the decay chain $\pi^{\pm} \rightarrow \mu^{\pm} \rightarrow e^{\pm}$, then there should be a significant flux of UHEGR resulting from the decays of π^0 mesons which are produced along with the charged pions (Cocconi 1959, Gould and Burbidge 1965). The predicted fluxes are, in fact, a few orders of magnitude higher than the values/upper limits observed experimentally (see Table 5.1), and it must be concluded, therefore, that the high energy electrons in the remnant are accelerated as such. The actual mechanism of acceleration of the electrons is as yet uncertain, but various suggestions have been made: Fermi acceleration in the magnetic irregularities in the remnant and production and injection by the pulsar associated with the SNR being favoured.

Although the existence of ultra-high energy electrons in the remnant is inferred by invoking synchrotron emission to explain the X-ray emission from the nebula, it is the inverse Compton scattering of high energy electrons against the ambient light which most effectively leads to the production of UHEGR. Several authors have calculated the expected UHEGR fluxes produced by this process; see Rieke and Weekes (1969) and references quoted therein. The predictions of UHEGR fluxes are uncertain by an order of magnitude because of the lack of precision of the inferred electron fluxes in the SNR which, in turn, results from the uncertainty in the magnetic field. As an example of the predictions we can take the calculations of Reike and Weekes (1969) who derived expected fluxes in the range 10^{-11}–10^{-10} cm^{-2} s^{-1} at $E_\gamma > 1$ TeV. The experimentally observed values/upper limits quoted in Table 5.1 for the CRAB nebula are seen to be generally lower than these predictions, and a likely explanation is that the magnetic field in the remnant is somewhat higher than the 3×10^{-4} G assumed by these and other authors. An alternative explanation is that part of the X-radiation from the CRAB nebula is resulting from inverse Compton scattering instead of from the synchrotron process.

5.4.2 Production of UHEGR in pulsar environs

Pulsars are believed to be rapidly rotating neutron stars possessing very large magnetic fields ($\sim 10^{12}$ G); see Smith (1977) and Manchester and Taylor (1977). It is attractive to postulate that such rapidly rotating oblique magnetic dipoles generate intense electric fields in which charged particles are accelerated to very high energies. Despite intense theoretical efforts, a self-consistent understanding of the pulsar magnetospheres and the processes going on there it has proved to be a very difficult proposition; see Michel (1982) for a recent review. This is so even in the case of radio emission by pulsars, where there is already considerable experimental information. The situation with respect to UHEGR emission, about which there exists only a limited body of experimental data at present, is even less clear. To our knowledge, there are no firm predictions published concerning the emission of UHEGR at $E_\gamma > 1$ TeV by pulsars.

Gunn and Ostriker (1969) have outlined a possible scenario in which electrons and protons are accelerated to very high energies near the velocity of light cylinder around a pulsar where the magnetic dipole moment makes an angle with its rotation axis; see also Ostriker (1972). According to these authors, the rotating magnetic dipole emits low frequency electromagnetic waves with the same frequency, Ω, as that of the pulsar. Since the gyro-frequency, $\omega_0 = eB/mc$, is very much larger than the wave frequency, Ω, particles injected in the wave zone find themselves in strong, nearly static, crossed electric and magnetic fields. In a time, very short compared to the period of the wave, the particles move relativistically in nearly the same direction as the wave's propagation vector. Thereafter they ride the wave at essentially a constant phase. The energy of the particle turns out to be independent of where it is injected provided it is far into the wave zone to begin with. According to the authors, the particle attains an energy E given by

$$E = 3.6 \times 10^{13} A(Z/A)^{2/3} \Omega_2^{4/3} (B_{s,12})^{2/3} R_6^2 \, \text{eV} \tag{5.13}$$

Here Z and A are the atomic number and atomic mass of the particle; Ω_2 is the neutron star angular velocity in units of 10^2 rad s^{-1}; $B_{s,12}$ is the surface field in units of 10^{12} G; and R_6 is the neutron star radius in units of 10^6 cm. For the CRAB pulsar, the above equation predicts energies of 2×10^{13} and 2×10^{14} eV for electrons and protons, respectively. The authors invoke the Goldreich–Julian theory (Goldreich and Julian 1969) and find that the rate of injection of particles is quite high. UHEGR could then be produced either by inverse Compton scattering of the electrons or through the decays of π^0 mesons produced in the nuclear interactions of protons with matter.

Goldreich and Julian (1969) have considered a pulsar model in which the dipole magnetic moment is aligned with the rotation axis. In this case, the open field lines extend beyond the light cylinder and close in a boundary zone near the SN shell. Charged particles escape along these lines and are electrostatically accelerated up to energies of

$$E_{max} = 3 \times 10^{12} Z R_6^3 B_{12} P^{-2} \text{ eV} \tag{5.14}$$

where Z is the charge of the particle; P is the pulsar period in seconds; and the other symbols are as defined above. It can be seen that in the case of the CRAB pulsar, particles can be accelerated to $\sim 3 \times 10^{15}$ eV.

Both the models just considered deal with the acceleration of charged particles at or beyond the light cylinder. On the other hand, Sturrock (1971), Ruderman and Sutherland (1975), and Massaro and Salvati (1979), among others, have considered models in which charged particles ($e^{\pm}$) are accelerated in the gaps near the polar caps across which electric potentials of $>10^{12}$ V have been calculated to exist. These models were developed largely to explain the various features observed in radio emission and low energy γ-rays, but high energy γ-rays could also result from curvature radiation emitted by these high energy electrons or through inverse Compton scattering. There is one difficulty, however, in getting UHEGR away from the pulsars in these models. In the regions where the γ-rays are produced, the magnetic fields are quite intense, approaching 10^{12} G, and in such intense fields, γ-rays undergo pair production even at energies as low as several MeV. Massaro and Salvati (1979) have shown that γ-rays of 100 GeV energy become attenuated by 3 to 4 orders of magnitude in the pulsar magnetospheres. In this context, the view expressed by Shabad and Usov (1982) is relevant. In the usual picture the curvature radiation photons are emitted tangentially to the line of force of magnetic field and propagate in straight-line paths. As they do so, the pitch angle between the photon wave vector and the magnetic field increases until a certain value when the photon is absorbed through pair production. Shabad and Usov (1982) point out that, if the resonant behaviour characteristic of vacuum polarisation in a magnetic field is considered, the above picture changes considerably. The trajectories of the γ-quanta are strongly bent towards the direction of the magnetic field (instead of being straight) provided that the latter is sufficiently intense. Consequently, the absorption of γ-rays by pair production on the magnetic field is suppressed.

Cheng *et al.* (1976) have considered particle acceleration in the outer gaps of the pulsar magnetospheres at distances ~0.6 times the light cylinder radius. At such distances, UHEGR absorption is considerably less because of the r^{-3} dependence of the magnetic field strength. For example, the Sturrock condition (equation (5.2)) tells us that in the case of the CRAB pulsar, attenuation is severe only at $E_\gamma > 10^{12}$ eV. The authors propose that γ-rays up to 10^{11} eV are produced by electrons accelerated in the outer gap.

Hinata (1977) has pointed out that if UHEGR at $E_\gamma > 1$ TeV are produced at distance $\gtrsim 36$ neutron star radii from the CRAB pulsar, through curvature radiation by electrons of $E_e \sim 10^{14}$ eV, the photons can escape pair production. The UHEGR intensities can be made to match those observed by the Mt. Hopkins group (Grindlay *et al.* 1976). The author has not presented details of how the electrons become accelerated to 10^{14} eV, however.

A number of authors (Milgrom and Pines 1978, Stepanian 1982, Vestrand and Eichler 1982, Eichler and Vestrand 1984, Hillas 1984) have presented various scenarios of UHEGR production by Cygnus X-3. All the authors assumed that Cygnus X-3 is a binary system consisting of a neutron star and a normal stellar companion. High energy charged particles emitted by the neutron star (see Figure 5.23) interact with the atmosphere of the companion star, thus producing the UHEGR one sees in the experiments. Since the generated photons are strongly beamed along the velocity vector of the incident particle, an observer detects only those photons that are produced when particles streaming towards him strike the intervening target, i.e. the stellar atmosphere. In this picture, the binary period (~4.8 h) is made to match the observed periodicity of the X-ray and UHEGR signals from Cygnus X-3. The radiations are expected to populate two distinct phase regions, separated by approximately $\Delta\phi = 0.5$. The problem of accelerating the charged particles is relegated to the neutron star, which incidentally has not been observed to date as a pulsar *per se* in any part of the electromagnetic spectrum. The failure to detect the pulsations from the pulsar is attributed to the absorption of radiations in the wind driven off the companion star by the young luminous pulsar. Stephens and Verma (1984) have pointed out that, in the magnetic fields ~1330 G believed to exist near the companion star, UHEGR with $E_\gamma > 10^{16}$ eV will be attenuated by pair production, thus leading to a steepening of the UHEGR spectrum at this energy, as observed.

5.4.3 Implications of UHEGR

The existence of finite fluxes of UHEGR has some interesting implications in the fields of cosmic ray physics and X-ray astronomy.

Wdowczyk and Wolfendale (1983, 1984) have noted that, for sources such as the CRAB pulsar/nebula and Cygnus X-3, the differential energy

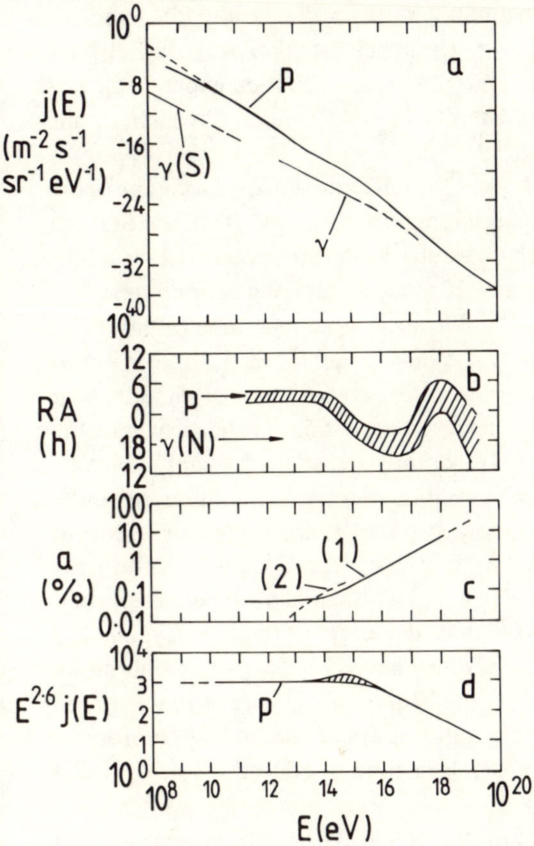

Figure 5.24. This figure, taken from Wdowczyk and Wolfendale (1983), shows in (*a*) the energy spectrum of cosmic ray particles, denoted *p*. γ(S): 10 % of the average γ-ray intensity with $|b| < 10°$ from satellite experiments. γ: the average γ-ray intensity with $|b| < 10°$ postulated by the authors. (*b*) Phase of first harmonic of cosmic ray anisotropy for the Northern Hemisphere. '*P*' is the postulated phase for protons and γ(*N*) that expected for γ-rays. (*c*) Amplitudes of first and second harmonics for Northern Hemisphere anisotropies. (*d*) Proton spectrum. Note that the change of slope occurs at a higher energy than the changes in (*b*) and (*c*). The shaded area represents (a somewhat contentious) 'bump' in the spectrum.

spectra of γ-rays seem to have flat exponents with $\gamma_d \sim -2$, while the cosmic ray differential energy spectrum has a steeper exponent of about -2.6 (up to a primary energy of $\sim 10^{15}$ eV). If all γ-ray sources in the Galaxy have $\gamma_d \sim -2$, then, normalising to the satellite γ-ray intensity from sources in the 100 MeV region, the γ-ray to cosmic ray proton ratio will increase with energy from about 3×10^{-6} at $E_\gamma \sim 100$ MeV to nearly 0.05 at 10^{15} eV; see Figure 5.24*a*. The resulting admixture of UHEGR with primary cosmic rays at energies $>10^{14}$ eV can then explain some of the hitherto inexplicable features of 'primary cosmic rays' which, in this picture, are not all charged particles.

In Figures 5.24*b* and *c* are shown the phase and the first harmonic of the primary cosmic ray anisotropy as a function of energy. It is clear that the phase of the first harmonic changes direction from about RA $\approx$ 3 h to about 20 h in the energy range $10^{15\pm1}$ eV, and the amplitude starts increasing abruptly at $E = 10^{14}$ eV from its constant value of $\approx$0.07 %. UHEGR then start to dominate the anisotropy for $E > 10^{14}$ eV in this model. In view of the fact that most of the cosmic ray experiments near 10^{14} eV are carried out at sea-level or mountain altitudes, where the nature of the primary is never identified, the model cannot yet be checked. If the γ/p ratio were to continue increasing with energy indefinitely then, above 10^{18} eV, γ-rays would predominate in the cosmic ray beam; however, based on the experience with Cygnus X-3, the γ-ray spectrum falls steeply above 10^{16} eV and a potential problem disappears. Wdowczyk and Wolfendale note that, if sources of the type Cygnus X-3 emit 10^{15} eV protons with the same efficiency as 10^{15} eV γ-rays, only about 30 Cygnus X-3's are needed in the Galaxy to generate the required intensity of cosmic ray particles; if the γ-rays are secondary to protons, however, the number of Cygnus X-3s needed is reduced to one or two at any one time.

Clay *et al.* (1984) have looked for an excess of air showers, in their Buckland Park air shower experiment, from the Galactic Plane and Galactic Centre which could be interpreted as due to UHEGR. Not finding any, the authors have placed an upper limit to the γ/p ratio at the 95 % confidence level of about 2 %. This upper limit is not in contradiction with the average value of ~1 % for $|b| < 10°$ suggested by Wdowczyk and Wolfendale (1983).

Rana *et al.* (1984) have looked at the consequences of γ-rays at $E_\gamma > 10^{15}$ eV propagating into the Galactic Halo. They argue that the UHEGR will interact with the 2.7 K microwave background photons, thereby producing electron–positron pairs, which in turn emit X-rays by synchrotron radiation in the Galactic magnetic field. This model predicts a flattening of the diffuse X-ray spectrum above 100 keV as measured at high Galactic latitudes even for only one Cygnus X-3 type source in the Galaxy. A firm detection of these hard X-rays would provide evidence for the γ–γ interaction and also confirm that the microwave radiation extends to at least 10 kpc from the Galactic Plane.

5.5 Summary

A number of experiments have shown that UHEGR are emitted by pulsars, such as the CRAB and VELA, by X-ray binaries typified by Cygnus X-3, Her X-1 and VELA X-1, and by external galaxies such as M31 and Cen A. The details of the results are given in Sections 5.2 and 5.3 and some of them are summarised in Tables 5.1 to 5.4. The situation can be summarised crudely in the following fashion. At $E_\gamma \sim 1$ TeV, the time-averaged UHEGR fluxes are $\sim 10^{-11}$ cm^{-2} s^{-1} to within an order of magnitude from most of the objects. The time-averaged luminosities at $E_\gamma > 10^{12}$ eV for Galactic objects are $\sim 3 \times 10^{33}$ erg s^{-1} to within a factor of 4, except in the case of Cygnus X-3 for which the luminosity is $\sim 2 \times 10^{37}$ erg s^{-1}. The luminosities for the extragalactic objects M31 and Cen A are quoted by the authors to be 4×10^{40} and 2×10^{41} erg s^{-1}. At $E_\gamma > 10^{15}$ eV, time-averaged UHEGR fluxes are in the region of 10^{-13} cm^{-2} s^{-1} to within a factor of 5; the time-averaged luminosities are $\sim 2.3 \times 10^{34}$ erg s^{-1} in the case of VELA X-1 and $\sim 10^{37}$ erg s^{-1} for the CRAB pulsar and Cygnus X-3.

All the results published so far have two properties in common: (a) poor statistics, and (b) time variability of source intensities. The detected signals are seldom larger than 4 or 5σ above the background, and the calculated probabilities for chance fluctuations in the background are in most cases not much below 10^{-3}. The same groups of authors using the same experimental set-up do not find signals every time and they therefore conclude that the source intensities vary with time, over time scales ranging from tens of milliseconds to years. Sceptics tend to deny the admissibility of the time-variability hypothesis, and think that whenever 'signals' are seen they are probably due to chance fluctuation in the background. The sceptics imply that the improbability calculations of such fluctuations made by the various authors probably did not take all the factors into account, and it must be admitted that such calculations are not easy to make.

It is unfortunate that, to our knowledge, no two experiments have collected data from the same object at the same time. Such simultaneous observations would be very significant in providing convincing arguments as to the reality of the effect. When, eventually, the signal strengths are much larger, say at the 10σ level or greater, it will be easier to establish the variability hypothesis.

Despite the sceptics' views, the present authors argue in favour of the experimenters in that when they use a deliberately wrong period in a periodicity analysis or when they look for excess counts from an uninteresting blank region in the sky or for statistical fluctuations on the negative side (i.e. deficiencies of more than 4σ), signals are not seen. A fair summary of the situation, then, is that at least some of the experiments have already detected UHEGR from celestial sources at just above the detection threshold.

In future, efforts must be made to lower the detection thresholds both in terms of energy and flux levels. Also, whenever possible, experiments should be coordinated to view the same object simultaneously. While the situation regarding which ephemerides should be used in the analysis of data from pulsars is satisfactory, the same is not true with respect to the important source Cygnus X-3. In the past, different experimenters have used different ephemerides derived from a variety of X-ray data. This has led to a confusion in the claimed phases, and even in the reality of the claim for detection of a pulsating source. There is merit in using a universally agreed-upon ephemeris by all the experimental groups. Prospects for obtaining an up-to-date and accurate ephemeris seem bright now with data streaming in from EXOSAT and the Japanese satellites.

As mentioned in Section 5.1, the subject of UHEGR has a strong bearing on the problem of the origin of cosmic rays. Even after the whole sky is searched for UHEGR sources and mapped in a statistically significant way many interesting problems will remain. It will be necessary to decide whether the γ-rays come from the interactions of protons via π^0 decay or from electrons (or positrons) via synchrotron emission, bremsstrahlung and inverse Compton scattering processes; a deduction of the intensities of these charged particles in the source regions will then be necessary. Another problem will be to consider the leakage of charged high energy particles into interstellar space and whether the lifetimes and numbers of sources are sufficient to account for the needed long term average energy input into the cosmic radiation. Finally, it will be necessary to account for the charge composition, energy spectrum and anisotropy of the cosmic radiation. The measurements described in this chapter are but a necessary first step towards solving the problem of the origin of cosmic radiation. With groups from Australia, Greece, India, Ireland, Japan, South Africa, the UK, USA and the USSR either planning to improve their existing capabilities to record UHEGR, or entering the field afresh, we can expect to witness a marked improvement in the quality and quantity of experimental information during the next decade.

Appendix

Gamma-ray sources

Object	Source direction: RA h min ° ′	Decl. ° ′	l (degrees)	b (degrees)	Distance (kpc)	Radio	Optical	X-ray	Gamma-rays: Gamma-ray lines	Medium high	(TeV)	(10^{15} eV)
CRAB Pulsar	05-32	+21-59	184.5	−5.8	2	○	○	○	×	○	○	○
VELA Pulsar	08-34	−45-00	263.6	−2.5	0.5	○	○	×		○	○	
Cygnus X-3	20-31	+40-47	79.9	+0.6	>11.4	○	×	○		○?	○	○*
Hercules X-1	16-56	+35-25	57.8	+37.7	5	×	○	○			○?	
VELA X-1	09-00	−40-21	263.0	+3.9	1.4	×	○	○		×		○?
Geminga	6-34	+18-00	195	+4	≈0.1	×	○	○		○	○?	
SS433	19-09	+04-54	39.7	−2.2	5.5	○	○	○	○		○?	
Galactic Centre	17-42	−28-54	0	0	≈10	○	○		○	○	○*	
M31 (Andromeda)	00-40	+41-00	121.4	−21.6	670		○				○?	
Centaurus A (NGC 5128)	13-22	−42-46	309.6	+19.4	6300	○	○	○			○	
3C 273	12-27	+02-20	289.3	+64.6	~9×10^5	○	○	○		○	×	×

○: observed at 4σ or higher significance
×: not seen
○?: claimed signals with less than 3.5σ effect. Needs confirmation
Blank: information not available
* decrease in intensity

REFERENCES

Chapter 1

Badhwar, G. D. and Stephens, S. A. (1977). *Proc. 15th Int. Cosmic Ray Conf.*, Plovdiv, **1**, 198.

Blandford, R. D. (1976). *Mon. Not. R. Astron. Soc.*, **176**, 465.

Chupp, E. L. (1976). *Gamma Ray Astronomy*, Reidel, Dordrecht.

Clayton, D. D., Colgate, S. A. and Fishman, D. J. (1969). *Astrophys. J.*, **155**, 75.

Dodds, D., Wolfendale, A. W. and Wdowczyk, J. (1976). *Mon. Not. Roy. Astr. Soc.*, **176**, 345.

Evans, R. D. (1955). *The Atomic Nucleus*, McGraw-Hill, New York.

Ginzburg, V. L. and Syrovatskii, S. I. (1964). *The Origin of Cosmic Rays*, Pergamon Press, New York.

Ginzburg, V. L. (1969). *Elementary Processes for Cosmic Ray Astrophysics*, Gordon and Breach, New York.

Hayakawa, S. (1969). *Cosmic Ray Physics*, Wiley-Interscience, New York.

Heitler, W. (1960). *The Quantum Theory of Radiation*, Oxford University Press, Oxford.

Hillier, R. (1984). *Gamma Ray Astronomy*, Oxford University Press, Oxford.

Jauch, J. M. and Rohrlich, F. (1955). *The Theory of Photons and Electrons*, Addison–Wesley, Cambridge, MA.

Lingenfelter, R. E., Higdon, J. C. and Ramaty, R. (1978). In *Gamma Ray Spectroscopy in Astrophysics* (Shen, B. S. P. ed.), p. 99, Benjamin, New York.

Lovelace, R. V. E. (1976). *Nature*, **262**, 649.

Marmier, P. and Sheldon, E. (1969). *Physics of Nuclei and Particles*, vol. I, Academic Press, New York.

Manchester, R. N. and Taylor, J. H. (1977). *Pulsars*, Freeman, San Francisco.

Meneguzzi, M. and Reeves, H. (1975). *Astron. Astrophys.*, **40**, 91.

Okeke, P. N. and Rees, M. J. (1980). *Astron. Astrophys.*, **81**, 263.

Ramaty, R. (1974). In *High Energy Particles and Quanta in Astrophysics* (McDonald, F. B. and Fichtel, C. E. eds), p. 122, MIT Press, Cambridge, MA.

Ramaty, R., Kozlovsky, B. and Lingenfelter, R. E. (1979). *Astrophys. J. Suppl.*, **40**, 487.

Ramaty, R. and Lingenfelter, R. E. (1981). *Phil. Trans. Roy. Soc. Lond.*, **A301**, 671.

Ramaty, R., Lingenfelter, R. E. and Kozlovsky, B. (1982). In *AIP Conf. Proc. 77* (Lingenfelter, R. E. *et al.* ed.), p. 211, AIP Press, New York.

Rossi, B. (1952). *High Energy Particles*, Prentice-Hall, Englewood Cliffs, NJ.

Stecker, F. W. (1971). *Cosmic Gamma Rays*, NASA SP-249 National Aeronautics and Space Administration, Washington DC.

Stecker, F. W. (1973). *Astrophys. & Space Sci.*, **20**, 47.

Stecker, F. W. (1975). In *Origin of Cosmic Rays* (Osborne, J. L. and Wolfendale, A. W. eds), p. 267, Reidel, Dordrecht.

Stephens, S. A. and Badhwar, G. D. (1981). *Astrophys. & Space Sci.*, **76**, 213.

Strong, A. W. and Wolfendale, A. W. (1981). *Phil. Trans. Roy. Soc. Lond.*, **A301**, 541.

Wdowczyk, J., Thackzyk, W. and Wolfendale, A. W. (1972). *J. Phys.*, **A5**, 1419.

Chapter 2

Abell, G. D. and Margon, B. (1979). *Nature*, **279**, 701.

Albernhe, F., Leborgne, J. F., Vedienne, G., Boclet, D., Durouchoux, P. and Da Coster, J. M. (1981). *Astron. Astrophys.*, **94**, 214.

Arnett, W. D. and Wefel, J. P. (1978). *Astrophys. J. Lett.*, **224**, L139.

Arnould, M., Norgaard, H., Thielemsn, F.-K. and Hillebrandt, W. (1980). *Astrophys. J.*, **237**, 931.

Begelman, M. C., Sarazin, C. L., Hatchett, S. P., McKee, C. F. and Arons, J. (1980). *Astrophys. J.*, **238**, 722.

Bhat, C. L., Houston, B. P., Issa, M. R., Mayer, C. J. and Wolfendale, A. W. (1985). *Nature*, **314**, 511.

Bussard, R. W., Ramaty, R. and Drachman, R. J. (1979). *Astrophys. J.*, **228**, 928.

Chupp, E. L. (1976). *Gamma Ray Astronomy*, Reidel, Dordrecht.

Clayton, D. D. (1973). In *Gamma-ray Astrophysics* (Stecker, F. W. and Trombka, J. eds), p. 263, NASA Spec. Publs. 339.

Clayton, D. D., Colgate, S. A. and Fishman, D. J. (1969). *Astrophys. J.*, **155**, 75.

Davidson, K. and McCray, R. (1980). *Astrophys. J.*, **241**, 1082.

Dean, A. J. and Ramsden, D. (1981). *Phil. Trans. Roy. Soc. Lond.*, **A301**, 577.

Fabian, A. C., Maccagni, D., Rees, M. J. and Stoeger, W. R. (1976). *Nature*, **260**, 683.

Graser, U. and Schonfelder, V. (1982). *Astrophys. J.*, **263**, 677.

Hall, R. D., Meegan, C. A., Walraven, G. D., Djuth, F. T. and Haymes, R. C. (1976). *Astrophys. J.*, **210**, 631.

Hemeury, J. M., Boclet, D., Durouchoux, P., Cline, T. L., Paciesas, W. S., Teegarden, B. J., Tueller, J. and Haymes, R. C. (1983). *Astrophys. J.*, **270**, 144.

Haymes, R. C., Walraven, G. D., Meegan, C. A., Hall, R. D., Djuth, F. T. and Shelton, D. M. (1975). *Astrophys. J.*, **210**, 593.

Hillebrandt, W. and Thielemann, F.-K. (1982). *Astrophys. J.*, **255**, 617.

Hillier, R. (1984). *Gamma-Ray Astronomy*, Clarendon Press, Oxford.

Johnson, W. N. and Haymes, R. C. (1973). *Astrophys. J.*, **184**, 103.

Lamb, R. C., Ling, J. C., Mahoney, W. A., Riegler, G. R., Wheaton, W. A. and Jacobson, A. S. (1983). *Nature*, **305**, 37.

Leventhal, M., MacCallum, C. J. and Watts, A. C. (1977). *Nature*, **266**, 696.

Leventhal, M., MacCallum, C. J. and Strong, P. D. (1978). *Astrophys. J. Lett.*, **225**, L11.

Leventhal, M., MacCallum, C. J., Huters, A. F. and Strong, P. D. (1980). *Astrophys. J.*, **240**, 338.

Leventhal, M., MacCallum, C. J., Huters, A. F. and Strong, P. D. (1982). *Astrophys. J. Lett.*, **260**, L1.

Ling, J. C., Mahoney, W. A., Willett, J. B. and Jacobson, A. S. (1979). *Astrophys. J.*, **231**, 896.

Lingenfelter, R. E., Higdon, J. C. and Ramaty, R. (1978). In *Gamma-Ray Spectroscopy in Astrophysics* (Cline, T. L. and Ramaty, R. eds), p. 252, NASA Tech. Mem. 79619.

Lingenfelter, R. E. and Ramaty, R. (1982). In *The Galactic Centre* (Riegler, G. R. and Blandford, R. D., eds), AIP Press, New York.

Mahoney, W. A., Ling, J. C., Jacobson, A. S. and Tapphorn, R. M. (1980). *Nucl. Instr. Meth.*, **178**, 363.

Mahoney, W. A., Ling, J. C., Jacobson, A. S. and Lingenfelter, R. E. (1982). *Astrophys. J.*, **262**, 742.

Mahoney, W. A., Ling, J. C. and Jacobson, A. S. (1984). *Astrophys. J.*, **278**, 784.

Margon, B. (1982). In *Accretion Driven Stellar X-Ray Sources* (Lewin, W. H. G. and Vanden Heuvel, E. P. T. eds), Cambridge University Press, Cambridge.

Marscher, A. P., Brecher, K., Wheaton, W. A., Ling, J. C., Mahoney, W. A. and Jacobson, A. S. (1984). *Astrophys. J.*, **281**, 566.

Matteson, J. L., Nolan, P. L. and Peterson, L. E. (1979). *X-Ray Astronomy*, p. 543, Pergamon Press, Oxford.

Milgrom, M. (1979). *Astron. Astrophys Lett.*, **76**, L3.

Paciesas, W. S., Cline, T. L., Teegarden, B. J., Tueller, J., Durouchoux, P. and Hameury, J. M. (1982). *Astrophys. J. Lett.*, **260**, L7.

Ramaty, R. and Lingenfelter, R. E. (1979). *Nature*, **278**, 127.

Ramaty, R. and Lingenfelter, R. E. (1981). *Phil. Trans. Roy. Soc. Lond.*, **A301**, 671.

Riegler, G. R., Ling, J. C., Mahoney, W. A. and Jacobson, A. S. (1983). *J.P.L. Astrophys.* Preprint 54, California Institute of Technology.

Riegler, G. R., Ling, J. C., Mahoney, W. A., Wheaton, W. A., Willett, J. B., Jacobson, A. S. and Prince, T. A. (1981). *Astrophys. J. Lett.*, **248**, L13.

Trombka, J. I. and Fichtel, C. E. (1982). NASA Tech. Mem. 84928.

Truran, J. W. and Cameron, A. G. W. (1978). *Astrophys. J.*, **219**, 226.

Woosley, S. E., Axelrod, T. A. and Weaver, T. A. (1981). *Comments Nucl. Particle Phys.*,

Woosley, S. E. and Weaver, T. A. (1980). *Astrophys. J.*, **238**, 1017.

Chapter 3

Aharonian, F. A., Aloyan, A. M. and Sunyaev, R. A. (1980). *Yerevan Phys. Inst. preprint 432(39)-80.*

Anderson, R. D., Peacock, A. and Taylor, B. G. (1984). *Adv. Space Res.*, **3**, (10–12), 485.

Barat, C., Chambon, G., Hurley, K., Niel, M., Vedrenne, G., Estulin, I. V., Kurt, V. G. and Zenchenko, V. M. (1979). *Astron. Astrophys. Lett.*, **79**, L24.

Barat, C., Chambon, G., Hurley, K., Niel, M., Vedrenne, G., Estulin, I. V., Kuznetsov, A. V. and Zenchenko, V. M. (1981). *Astrophys. & Space Sci.*, **75**, 83.

Barat, C., Chambon, G., Hurley, K., Niel, M. and Vedrenne, G. (1982). *Astron. Astrophys. Lett.*, **109**, L9.

Barat, C., Hayles, R. I., Hurley, K., Niel, M., Vedrenne, G., Desai, U., Estulin, I. V., Kurt, V. G. and Zenchenko, V. M. (1983). *Astron. Astrophys.*, **126**, 400.

Barat, C., Hayles, R. J., Hurley, K., Niel, M., Vedrenne, G., Estulin, I. V. and Zenchenko, V. M. (1984*a*). *Astrophys. J.*, **285**, 791.

Barat, C., Hurley, K., Niel, M., Vedrenne, G., Evans, W. D., Fenimore, E. E., Klebesadel, R. W., Laros, J. G., Cline, T. L., Estulin, I. V., Zenchenko, V. M. and Kurt, V. G. (1984*b*). *Astrophys. J.*, **280**, 150.

Barat, C., Niel, M., Hayles, R. I., Hurley, K., Vedrenne, G., Mitrofanov, I. G., Estulin, I. V., Zenchenko, V. M. and Dolidze, V. S. (1984*c*). *Adv. Space Res.*, **3** (10–12), 105.

Barat, C., Hurley, K., Niel, M., Vedrenne, G., Cline, T., Desai, U., Schaefer, B., Teegarden, B., Evans, W. D., Fenimore, E. E., Klebesadel, R., Laros, G., Estulin, I. V., Zenchenko, V. M., Kuznetsov, A. V., Kurt, V. G., Ilovaisky, S. and Motch, C. (1984*d*). *Astrophys. J. Lett.*, **286**, L5.

Beurle, K., Bewick, A., Mills, J. S. and Quenby, J. J. (1981). *Astrophys. & Space Sci.*, **77**, 201.

Bewick, A., Coe, M. J., Quenby, J. J. and Mills, J. S. (1975). *Nature*, **258**, 686.

Bhat, P. N., Gopalakrishnan, N. V., Gupta, S. K., Ramana, Murthy, P. V., Sreekantan, B. V., Tonwar, S. C. and Vishwanath, P. R. (1982). *Mon. Not. Roy. Astron. Soc.*, **199**, 1007.

Bisnovatyi-Kogan, G. S. and Chechetkin, V. M. (1980). Preprint 561, Inst. for Space Res., Acad. Nauk., USSR.

Bonazzola, S., Hameury, J. M., Heyvaerts, J. and Lasota, J. P. (1984). *Astron. Astrophys.*, **136**, 89.

Brecher, K. (1982). In *AIP Conf. Proc. 77* (Lingenfelter, R. E. *et al.* eds), p. 293, AIP Press, New York.

Bussard, R. W. and Lamb, F. K. (1982). In *AIP Conf. Proc. 77* (Lingenfelter, R. E. *et al.* eds), p. 189, AIP Press, New York.

Cavallo, G. and Rees, M. J. (1978). *Mon. Not. Roy. Astron. Soc.*, **183**, 359.

Chambon, G. (1982). Doctoral Thesis 1045, UPS, Toulouse, France.

Cline, T. L. (1980). *Comments on Astrophys.*, **9**, 13.

Cline, T. L. (1982). In *AIP Conf. Proc. 77* (Lingenfelter, R. E. *et al.* eds), p. 17, AIP Press, New York.

Cline, T. L. (1983). Adv. Space Res., **3**, (4), 175.

Cline, T. L., Desai, U. D., Klebesadel, R. W. and Strong, I. B. (1973). *Astrophys. J. Lett.*, **185**, L1.

Cline, T. L. and Desai, U. D. (1974). *Proc. ESLAB Symposium*, **37**, ESRO, Noordwijk.

Cline, T. L. and Desai, U. D. (1975). *Astrophys. J. Lett.*, **196**, L43.

Cline, T. L., Desai, U. D., Schmidt, W. K. H. and Teegarden, B. J. (1977). *Nature*, **266**, 694.

Cline, T. L., Desai, U. D.and Teegarden, B. J. (1981*a*). *Astrophys. & Space Sci.*, **75**, 93.

Cline, T. L., Desai, U. D., Pizzichini, G., Teegarden, B. J., Evans, W. D., Klebesadel, R. W., Laros, J. G., Barat, C., Hurley, K., Niel, M., Vedrenne, G., Estulin, I. V., Mersov, G. A., Zenchenko, V. M. and Kurt, V. G. (1981*b*). *Astrophys. J. Lett.*, **246**, L133.

Cline, T. L., Desai, U. D., Teegarden, B. J., Evans, W. D., Klebesadel, R. W., Laros, J. G., Barat, C., Hurley, K., Niel, M., Vedrenne, G., Estulin, I. V., Kurt, V. G., Mersov, G. A., Zenchenko, V. M., Weisskopf, M. C. and Grindlay, J. (1982). *Astrophys. J. Lett.*, **255**, L45.

Colgate, S. A. (1968). *Can. J. Phys.*, **46**, S476.

Colgate, S. A. (1982). In *AIP Conf. Proc. 77* (Lingenfelter, R. E. *et al.* eds), p. 309, AIP Press, New York.

Colgate, S. A. and Petschek, A. G. (1981). *Astrophys. J.*, **248**, 771.

Daugherty, J. K. and Bussard, R. W. (1980). *Astrophys. J.*, **238**, 296.

Dennis, B. R., Frost, K. J., Kiplinger, A. L., Orwig, L. E., Desai, U. and Cline, T. L. (1982).

Evans, W. D., Belian, R. D. and Conner, J. P. (1976). *Astrophys. J. Lett.*, **207**, L91.

Evans, W. D., Klebesadel, R. W., Laros, J. G., Cline, T. L., Desai, U. D., Pizzichini, G., Teegarden, B. J., Hurley, K., Niel, M., Vedrenne, G., Estoolin, I. V., Kowznetsov, A. V., Zenchenko, V. M. and Kurt, V. G. (1980). *Astrophys. J. Lett.*, **237**, L7.

Evans, W. D., Fenimore, E. E., Klebesadel, R. W., Laros, J. G. and Terrell, N. J. (1981). *Astrophys. & Space Sci.*, **75**, 35.

Fabian, A. C., Icke, V. and Pringle, J. E. (1972). *Astrophys. & Space Sci.*, **42**, 77.

Felten, J. E. (1981). *Proc. 17th Int. Cosmic Ray Conf.*, Paris, **9**, 52.

Fenimore, E. E., Klebesadel, R. W., Laros, J. G. and Stockdale, R. E. (1982*a*). *Nature*, **297**, 665.

Fenimore, E. E., Laros, J. G., Klebesadel, R. W., Stockdale, R. E. and Kane, S. R. (1982*b*). In *AIP Conf. Proc. 77* (Lingenfelter, R. E. *et al.* eds), p. 201, AIP Press, New York.

Fishman, G. J. (1979). *Astrophys. J.*, **233**, 851.

Fishman, G. J. (1981). *Astrophys. & Space Sci.*, **75**, 125.

Fishman, G. J., Meegan, C. A., Watts, J. W. and Derrickson, J. H. (1978). *Astrophys. J. Lett.*, **223**, L13.

Fishman, G. J., Meegan, C. A., Parnell, T. A. and Wilson, R. B. (1982). In *AIP Conf. Proc. 77* (Lingenfelter, R. E. *et al.* eds), p. 443, AIP Press, New York.

Forrest, D. J., Chupp, E. L., Ryan, J. M., Cherry, M. L., Gleske, I. U., Reppin, C., Pinkau, K., Rieger, E., Kanbach, G., Kinzer, R. L., Share, G. H., Johnson, W. N. and Kurfess, J. D. (1980). *Solar Phys.*, **65**, 15.

Gilman, D., Metzer, A. E., Parker, R. H., Evans, L. and Trombka, J. I. (1980). *Astrophys. J.*, **236**, 951.

Golenetskii, S. V. *et al.* (1979). *Sov. Astron. Lett.*, **5**, 340.

Golenetskii, S. V., Mazets, E. P., Aptekar, R. L. and Ilyinski, V. N. (1983). *Nature*, **306**, 451.

Golenetskii, S. V., Ilyinskii, V. N. and Mazets, E. P. (1984). *Nature*, **307**, 41.

Grindlay, J. E., Wright, E. L. and McCrosky, R. E. (1974). *Astrophys. J. Lett.*, **192**, L113.

Grindlay, J. E., Cline, T., Desai, U. D., Teegarden, B. J., Pizzichini, G., Evans, W. D., Klebesadel, R. W., Laros, J. G., Hurley, K., Niel, M. and Vedrenne, G. (1983). *Nature*, **300**, 730.

Harwit, M. and Salpeter, E. E. (1973). *Astrophys. J. Lett.*, **186**, L37.

Helfand, D. J. and Long, K. S. (1979). *Nature*, **282**, 589.

Helfand, D. J. and Vrtilek, S. D. (1983). *Nature*, **304**, 41.

Hewish, A., Bell, S. J., Pilkington, J. D. H., Scott, P. F. and Collins, R. A. (1968). *Nature*, **217**, 709.

Hjellming, R. and Ewald, S. P. (1981). *Astrophys. J. Lett.*, **246**, L137.

Hjellming, R. (1983). Private communication to Barat *et al.* (1984*b*).

Hudec, R., Ceplecha, Z., Ehrlich, J., Borovicka, J., Hurley, K., Ateia, J. L., Barat, C., Niel, M., Vedrenne, G., Estulin, I. V., Kuznetsov, A. K., Zenchenko, V. M., Cline, T. L., Desai, U., Evans, W. D., Fenimore, E., Klebesadel, R. and Laros, J. (1984). *Adv. Space Res.*, **3**, (10), 115.

Hueter, G. J. and Gruber, D. E. (1982). In *Proc. Workshop on Accreting Neutron Stars* (Brinkmann, W. and Trumper, J. eds), Max Planck Institute Report 177, p. 213.

Hurley, K. (1982). In *AIP Conf. Proc. 77* (Lingenfelter, R. E. *et al.* eds), p. 85, AIP Press, New York.

Hurley, K. (1983*a*). *Adv. Space Res.*, **3**, (4), 163.

Hurley, K. (1983*b*). In *AIP Conf. Proc. 101* (Burns, M. L. *et al.* eds), p. 21, AIP Press, New York.

Hurley, K. (1983*c*). *Adv. Space Res.*, **3**, (4), 203.

Hurley, K. (1984). In *AIP Conf. Proc. 115* (Woosely, S. E. ed.), p. 10, AIP Press, New York.

Jennings, M. C. (1982*a*). *Astrophys. J.*, **258**, 110.

Jennings, M. C. (1982*b*). In *AIP Conf. Proc. 77* (Lingenfelter, R. E. *et al.* eds), p. 107, AIP Press, New York.

Jennings, M. C. and White, R. S. (1980). *Astrophys. J.*, **238**, 110.

Joss, P. C. (1978). *Astrophys. J. Lett.*, **225**, L123.

Joss, P. C. (1979). *Comments Astrophys.*, **8**, 109.

Katz, I. (1981). As quoted by Teegarden (1982).

Katz, I. (1982). *Astrophys. J.*, **260**, 371.

Katz, I. (1983). In *AIP Conf. Proc. 101* (Burns, M. L. *et al.* eds), p. 65, AIP Press, New York.

Klebesadel, R. W., Strong, I. B. and Olson, R. A. (1973). *Astrophys. J. Lett.*, **182**, L85.

Klebesadel, R. W., Evans, W. D. and Laros, J. G. (1981). *Astrophys. & Space Sci.*, **75**, 5.

Klebesadel, R. W., Fenimore, E. E. and Laros, J. (1984). In *AIP Conf. Proc. 115* (Woosely, S. E. ed.), p. 20, AIP Press, New York.

Knight, F. K., Matteson, J. L. and Peterson, L. E. (1981). *Astrophys. & Space Sci.*, **75**, 21.

Lamb, D. Q. (1982). In *AIP Conf. Proc. 77* (Lingenfelter, R. E. *et al.* eds), p. 249, AIP Press, New York.

Lamb, D. Q., Lamb, F. K. and Pines, D. (1973). *Nature Phys. Sci.*, **246**, 52.

Laros, J. G., Evans, W. D., Fenimore, E. E., Klebesadel, R. W., Barat, C., Hurley, K., Niel, M., Vedrenne, G., Estulin, I. V., Zenchenko, V. M. and Mersov, G. A. (1981). *Astrophys. J. Lett.*, **245**, L63.

Laros, J. G., Evans, W. D., Fenimore, E. E. and Klebesadel, R. W. (1982). *Astrophys. & Space Sci.*, **88**, 243.

Lasota, J. P. and Belli, B. M. (1983). *Nature*, **304**, 139.

Liang, E. P. (1981). *Nature*, **292**, 319.

Liang, E. P. (1982). *Nature*, **299**, 321.

Liang, E. P., Jerigan, T. E. and Rodrigues, R. (1983). *Ap. J.*, **271**, 766.

Ling, J. C., Mahoney, W. A., Willet, J. B. and Jacobson, A. S. (1982). In *AIP Conf. Proc. 77* (Lingenfelter, R. E. *et al.* eds), p. 143, AIP Press, New York.

Lingenfelter, R. E., Higdon, J. C. and Ramaty, R. (1978). In *Gamma Ray Spectroscopy in Astrophysics* (Cline, T. L. and Ramaty, R. eds), p. 252, NASA, T.M. 79619.

London, R. A. and Cominsky, L. R. (1983). *Astrophys. J. Lett.*, **275**, L59.

Loznikov, V. and Kuznetsov, A. V. (1982). *Acad. Sci. USSR*, preprint 743.

Mathewson, D. W. and Clark, J. N. (1973). *Astrophys. J.*, **180**, 725.

Mazets, E. P. (1984). Private communication to Vedrenne (1984).

Mazets, E. P., Golenitski, S. V., Ilyinski, V. N., Aptekar, R. L. and Guryan, Y. A. (1979*a*). *Pisma Astron. Zh.*, **5**, 307 (*Soviet Astron. Lett.*, **5**, 163 (1980)).

Mazets, E. P., Golenitskii, S. V., Ilyinski, V. N., Aptekar, R. L. and Guryan, Y. A. (1979*b*). *Nature*, **282**, 587.

Mazets, E. P. *et al.* (1979*c*). *Soviet Astron. Lett.*, **5**, 343.

Mazets, E. P., Golenitskii, S. V., Aptekar, R. L., Guryan, Y. A. and Ilyinski, V. N. (1980). *Pisma Astron. Zh*, **6**, 706.

Mazets, E. P. and Golenitskii, S. V. (1981*a*). *Astrophys. & Space Phys. Rev.*, **1**, 205.

Mazets, E. P. and Golenitskii, S. V. (1981*b*). *Astrophys. & Space Sci.*, **75**, 47.

Mazets, E. P. *et al.* (1981*a*). Preprint 738, A. F. Ioffe Physical-Technical Inst., Leningrad, USSR.

Mazets, E. P., Golenitskii, S. V., Ilyinski, V. N., Panov, V. N., Aptekar, R. L., Guryan, Y. A., Proskura, M. P., Sokolov, I. A., Sokolova, Z. A. and Kharitonova, T. V. (1981*b*). *Astrophys. & Space Sci.*, **80**, 3–83, 85–117, 119–43.

Mazets, E. P. *et al.* (1981*c*). Preprint 719, A. F. Ioffe Physical-Technical Inst., Leningrad, USSR.

Mazets, E. P., Golenitskii, S. V., Aptekar, R. L., Guryan, Y. A. and Ilyinski, V. N. (1981*d*). *Nature*, **290**, 378.

Mazets, E. P., Golenitskii, S. V., Ilyinski, V. N., Guryan, Y. A., Aptekar, R. L., Panov, V. N., Sokolov, I. A., Sokolova, Z. Y. and Kharitonova, T. V. (1982*a*). *Astrophys. & Space Sci.*, **82**, 261.

Mazets, E. P., Golenitskii, S. V., Guryan, Y. A. and Ilyinski, V. N. (1982*b*). *Astrophys. & Space Sci.*, **84**, 173.

Mazets, E. P., Golenitskii, S. V., Guryan, Y. A., Aptekar, R. L., Ilyinski, V. N. and Panov, V. N. (1983). In *AIP Conf. Proc. 101* (Burns, M. L. *et al.* eds), p. 36, AIP Press, New York.

Metzger, A. E., Parker, R. H., Gilman, D., Peterson, L. E. and Trombka, J. I. (1974). *Astrophys. J. Lett.*, **194**, L19.

Mitrofanov, I. G., Estulin, I. V., Zenchenko, V. M., Dolidze, V. S., Barat, C., Niel, M., Hayles, R. I., Hurley, K. and Vedrenne, G. (1984). *Adv. Space Res.*, **3**, (10–12), 119.

Newman, M. J. and Cox, A. N. (1980). *Astrophys. J.*, **242**, 319.

Norris, J. P., Cline, T. L., Desai, U. D. and Teegarden, B. J. (1984). *Nature*, **308**, 434.

Pedersen, H., Motch, C., Tarenghi, M., Danziger, J., Pizzichini, G. and Lewin, W. H. G. (1983). *Astrophys. J. Lett.*, **270**, L43.

Pedersen, H., Danziger, J., Hurley, K., Pizzichini, G., Motch, C., Ilovaisky, S., Gradmann, N., Brinkmann, W., Kanbach, G., Rieger, E., Reppin, C., Trumper, W. and Lund, N. (1984). *Nature*, **312**, 46.

Petrosian, V. (1981). *Astrophys. J.*, **251**, 727.

Pizzichini, G. (1981). *Astrophys. & Space Sci.*, **75**, 205.

Prilutskii, O. P., Rozental, I. L. and Usov, V. V. (1975). *Usp. Fiz. Nauk.*, **116**, 517 (*Soviet Phys. Usp.* **18**, 548, 1976).

Ramaty, R., Bonazzola, S., Cline, T. L., Kazanas, D., Metzaros, P. and Lingenfelter, R. E. (1980). *Nature*, **287**, 122.

Ramaty, R., Lingenfelter, R. E. and Bussard, R. W. (1981). *Astrophys. & Space Sci.*, **75**, 193.

Ramaty, R., McKinley, J. M. and Jones, F. C. (1982).

Rieger, E., Reppin, C., Kanbach, G., Forrest, D. J., Chupp, E. L. and Share, G. H. (1982). In *Proc. Workshop on Accreting Neutron Stars* (Brinkmann, W. and Trumper, J. eds), Max Planck Inst. report 177, p. 229.

Rothschild, R. E. and Lingenfelter, R. E. (1984). *Nature*, **312**, 737.

Ruderman, M. (1975). *Ann. N.Y. Acad. Sci.*, **262**, 164.

Ruderman, M. (1981). *Prog. Particle and Nucl. Phys.* **6**, 215 (Wilkinson, D. ed.), Pergamon Press, Oxford.

Schaefer, B. E. (1981). *Nature*, **294**, 722.

Schaefer, B. E. and Ricker, G. R. (1983). *Nature*, **302**, 43.

Schaefer, B. E., Seitzer, P. and Bradt, H. V. (1983). *Astrophys. J. Lett.*, **270**, L49.

Schaefer, B. E., Bradt, H. V., Barat, C., Hurley, K., Niel, M., Vedrenne, G., Cline, T. L., Desai, U., Teegarden, B. J., Evans, W. D., Fenimore, E. E., Klebesadel, R. W., Laros, J. G., Estulin, I. V. and Kuznetsov, A. V. (1984). *Astrophys. J. Lett.*, **286**, L1.

Schmidt, W. K. H. (1978). *Nature*, **271**, 525.

Strong, I. B. *et al.* (1973). Los Alamos Conf. on Transient Cosmic Gamma and X-ray sources, unpublished.

Teegarden, B. J. and Cline, T. L. (1980). *Astrophys. J. Lett.*, **236**, L67.

Teegarden, B. J. and Cline, T. L. (1981). *Astrophys. & Space Sci.*, **75**, 181.

Teegarden, B. J. (1982). In *AIP Conf. Proc. 77* (Lingenfelter, R. E. *et al.* eds), p. 123, AIP Press, New York.

Terrell, J., Evans, W. D., Klebesadel, R. W. and Laros, J. D. (1980). *Nature*, **285**, 383.

Terrell, J., Fenimore, E. E., Klebesadel, R. W. and Desai, U. D. (1982). *Astrophys. J.*, **254**, 279.

Trumper, J. (1978). In *Gamma Ray Spectroscopy in Astrophysics*, NASA Technical Memorandum 79619.

Tsygan, A. I. (1975). *Astrophys. & Space Sci.*, **44**, 21.

Tyson, J. and Jarvis, J. (1979). *Astrophys. J. Lett.*, **230**, L153.

Van Buren, D. (1982). In *AIP Conf. Proc. 77* (Lingenfelter, R. E. *et al.* eds), p. 311, AIP Press, New York.

Van Horn, H. M. and Hansen, C. J. (1974). *Astrophys. J.*, **191**, 479.

Vedrenne, G. (1983). Paper presented at the Symposium on Problems of Collapse and Numerical Relativity, Toulouse, France.

Vedrenne, G. (1984). *Adv. Space Res.*, **3**,(10), 97.

Vedrenne, G. and Chambon, G. (1983). *Space Sci. Rev.*, **36**, 319.
Ventura, J., Bonazzola, S., Hameury, J. M. and Heyvaerts, J. (1983). *Nature*, **301**, 491.
Vetter, F. (1982). *Phys. Rep.* **81**, 293.
Wood, K. S., Byram, E. T., Chubb, T. A., Friedman, H., Meekins, J. F., Share, G. H. and Yentis, J. A. (1981). *Astrophys. J.*, **247**, 632.
Woosley, S. E. (1982). In *AIP Conf. Proc. 77* (Lingenfelter, R. E. *et al.* eds), p. 273, AIP Press, New York.
Woosley, S. E. and Wallace, R. K. (1982). *Astrophys. J.*, **258**, 716.
Yoshimori, M. (1978). *Austr. J. Phys.*, **31**, 189.

Chapter 4

Apparao, K. M. and Hoffman, J. (1970). *Ap. J.*, **5**, 25.
Baker, R. E., Bassani, L., Dean, A. J., Ramsden, D., Butler, R. C., di Cocco, G., Boella, G., Della Ventura, A., Perotti, F. and Villa, G. (1981). *Proc. 17th Int. Cosmic Ray Conf.*, Paris, **1**, 222.
Bania, T. M. (1977). *Astrophys. J.*, **216**, 381.
Bennett, K., Bignami, G. F., Boella, G., Buccheri, R., Burger, J. J., Cuccia, A., Hermsen, W., Higdon, J., Kanbach, G., Koch, L., Lichti, G. G., Masnou, J., Mayer-Hasselwander, H. A., Paul, J. A., Scarsi, L., Shukla, P. G., Swanenburg, B. N., Taylor, B. G. and Wills, R. D. (1976). *The Structure and Content of the Galaxy and Galactic Gamma Rays*, Goddard Space Flight Centre X-662-76-154, 23.
Bennett, K., Bignami, G. F., Boella, G., Buccheri, R., Hermsen, W., Kanbach, G., Lichti, G. G., Masnou, J. L., Mayer-Hasselwander, H. A., Paul, J. A., Scarsi, L., Swanenburg, B. N., Taylor, B. G. and Wills, R. D. (1977). *Astron. Astrophys.*, **61**, 279.
Berkhuijsen, E. M., Haslam, C. G. T. and Salter, C. J. (1971). *Astron. Astrophys.*, **14**, 242.
Bhat, C. L., Mayer, C. J. and Wolfendale, A. W. (1984*a*). *Astron. Astrophys.*, **140**, 284.
Bhat, C. L., Issa, M. R., Houston, B. P., Mayer, C. J. and Wolfendale, A. W. (1984*b*). RAL Workshop on Astron. and Astrophys., RAL-84-101 (Gondhalekar, P. M. ed.), 39.
Bhat, C. L., Issa, M. R., Houston, B. P., Mayer, C. J. and Wolfendale, A. W. (1985*a*). *Nature*, **314**, 511.
Bhat, C. L., Issa, M. R., Mayer, C. J. and Wolfendale, A. W. (1985*b*). *Nature*, **314**, 515.
Bhat, C. L., Issa, M. R., Mayer, C. J. and Wolfendale, A. W. (1986) (in preparation).
Bignami, G. F., Fichtel, C. E., Hartman, R. C. and Thompson, D. J. (1979). *Astrophys. J.*, **232**, 649.
Bignami, G. F., Bennett, K., Buccheri, R., Caraveo, P. A., Hermsen, W., Kanbach, G. G., Lichti, G. G., Masnou, J. L., Mayer-Hasselwander, H. A., Paul, J. A., Sacco, B., Scarsi, L., Swanenburg, B. N. and Wills, R. D. (1981). *Astron. Astrophys.*, **93**, 71.
Bignami, G. F. and Hermsen, W. (1983). *Ann. Rev. Astron. Astrophys.*, **211**, 67.
Bignami, G. F., Caraveo, P. A. and Paul, J. A. (1984). *Nature*, **310**, 464.
Bignami, G. F. (1985). In *Cosmic Radiation in Contemporary Astrophysics*, International School of Cosmic Ray Astrophysics (Shapiro, M. M. ed.), Reidel, Dordrecht (in press).
Blandford, R. D. and Cowie, L. L. (1982). *Astrophys. J.*, **260**, 625.
Blitz, L. and Shu, F. H. (1980). *Astrophys. J.*, **238**, 148.
Blitz, L. (1980). In *Giant Molecular Clouds in the Galaxy* (Solomon, P. M. and Edmunds, M. G. eds), Pergamon Press.
Bloemen, J. B. G. M., Bennett, K., Bignami, G. F., Blitz, L., Caraveo, P. A., Gottwald, M., Hermsen, W., Lebrun, F., Mayer-Hasselwander, H. A. and Strong, A. W. (1984*a*). *Astron. Astrophys.*, **135**, 12.
Bloemen, J. B. G. M., Caraveo, P. A., Hermsen, W., Lebrun, F., Maddalena, R. J., Strong, A. W. and Thaddeus, P. (1984*b*). *Astron. Astrophys.*, **139**, 37.
Brown, R. W. and Stecker, F. W. (1979). *Phys. Rev. Lett.*, **43**, 315.
Buccheri, R. (1980). *Advances in Space Exploration*, vol. 7, p. 17, Pergamon Press, Oxford.

Buccheri, R., Morini, M. and Sacco, B. (1981). *Phil. Trans. Roy. Soc. Lond.*, **A301**, 495.

Caraveo, P. A., Bennett, K., Bignami, G. F., Hermsen, W., Kanbach, G., Lebrun, F., Masnou, J. L., Mayer-Hasselwander, H. A., Paul, J. A., Sacco, B., Scarsi, L., Strong, A. W., Swanenburg, B. N. and Wills, R. D. (1980). *Astron. Astrophys. Lett.*, **91**, L3.

Caraveo, P. A. (1981). *Phil. Trans. Roy. Soc. Lond.*, **A301**, 569.

Caraveo, P. A., Bignami, G. F., Vigroux, L. and Paul, J. A. (1984). *Ap. J. Lett.*, **276**, L45.

Cavaliere, A., Morrison, P. and Pacini, F. (1970). *Ap. J. Lett.*, **162**, L133.

Clark, G. W., Garmire, G. P. and Kraushaar, W. L. (1968). *Astrophys. J. Lett.*, **153**, L203.

Clark, G. W. (1971). IAU Symp. 41 (La Buhn, F. and Lust, R. eds), p. 3, Reidel, Dordrecht.

Dean, A. J. and Ramsden, D. (1981). *Phil. Trans. Roy. Soc.*, **A301**, 577.

Derdyn, S. M., Ehrmann, C. H., Fichtel, C. E., Kniffen, D. A. and Ross, R. W. (1972). *Nuc. Inst. & Meth.*, **98**, 557.

Dodds, D., Strong, A. W., Wolfendale, A. W. and Wdowczyk, J. (1974). *Nature*, **250**, 716.

Dodds, D., Strong, A. W. and Wolfendale, A. W. (1975). *Mon. Not. Roy. Astr. Soc.*, **171**, 569.

Eaton, N., Adams, D. J. and Giles, A. N. (1984). *Mon. Not. Roy. Astr. Soc.*, **208**, 241.

Fabian, A. C. and Nulsen, P. E. J. (1984). *Nature*, **312**, 48.

Fazio, G. G. (1970). *Nature*, **225**, 905.

Fazio, G. G. (1973). In *X-Ray and Gamma Ray Astronomy*, Proc. IAU Symp 55 (Madrid) (Bradt, H. and Giacconi, R. eds), Reidel, Dordrecht.

Fichtel, C. E., Hartman, R. C., Kniffen, D. A. and Sommer, M. (1972). *Astrophys. J.*, **171**, 31.

Fichtel, C. E., Hartman, R. C., Kniffen, D. A., Thompson, D. J., Bignami, G. F., Ogelman, H., Ozel, M. E. and Tumer, T. (1975). *Astrophys. J.*, **198**, 163.

Fichtel, C. E., Hartman, R. C., Kniffen, D. A., Thompson, D. J., Bignami, G. F., Ogelman, H., Ozel, M. E. and Tumer, T. (1975). *Astrophys. J.*, **198**, 163.

Fichtel, C. E., Hartman, R. C., Kniffen, D. A., Thompson, D. J., Ogelman, H., Ozel, M. E. and Tumer, T. (1977). *Astrophys. J. Lett.*, **217**, L9.

Fichtel, C. E., Hartman, R. C., Kniffen, D. A., Thompson, D. J., Ogelman, H., Tumer, T. and Ozel, M. E. (1978). NASA Tech. Memo. 79650.

Fichtel, C. E., Simpson, G. A. and Thompson, D. J. (1978). *Astrophys. J.*, **222**, 833.

Fichtel, C. E. (1981). *Phil. Trans. Roy. Soc. Lond.*, **A301**, 692.

Fichtel, C. E. (1982). NASA Technical Memo. 83957.

Fichtel, C. E. (1983). In *Advances in Space Research* (Vedrenne, G. and Hurley, K. Eds), Vol. 3, no. 4, p. 5, Pergamon Press, Oxford.

Galper, A. M., Kirillov-Ugryumov, U. G., Kurochkin, A. V., Leikov, N. G., Luchkov, B. I., Hurkin, Uy. T., Fomin, V. P., Neshpor, Yu. I., Stepanian, A. A. and Vladimirsky, B. M. (1977). *Proc. 15th Int. Cosmic Ray Conf.*, Plovdiv, **1**, 131.

Ginzburg, V. L. (1968). *Astrophys. & Space Sci.*, **1**, 1.

Gordon, M. A. and Burton, W. B. (1976). *Astrophys. J.*, **208**, 346.

Goned, A. (1981). *Proc. 17th Int. Cosmic Ray Conf.*, Paris, **1**, 210.

Hartman, R. C., Fichtel, C. E., Kniffen, D. A., Lamb, R. C., Thompson, D. J., Bignami, G. F., Ogelman, H., Ozel, M. and Tumer, T. (1976). *The Structure and Content of the Galaxy and Galactic Gamma Rays*, p. 12, Goddard Space Flight Centre, X-662-76-154.

Haslam, C. G. T., Kearsey, S., Osborne, J. L., Phillips, S. and Stoffel, H. (1981). *Nature*, **289**, 470.

Hayakawa, S. (1952). *Prog. Theor. Phys.*, **8**, 517.

Heiligman, G. M. (1982). Ph.D. Thesis, Princeton University.

Henrichs, H. F. and Van den Heuvel, E. P. J. (1983). *Nature*, **303**, 213.

Hermsen, W. (1980). Ph.D. Thesis, University of Leiden.

Hermsen, W. (1981). *Phil. Trans. Roy. Soc. Lond.*, **A401**, 519.

Hermsen, W. and Bloemen, J. B. G. M. (1982). *Astrophysics and Space Science Library*, Reidel, Dordrecht.

Hermsen, W. and Bloemen, J. B. G. M. (1983). *IAU Symp.*, **106**, 65.

Hess, V. F. (1912). *Phys. Z*, **13**, 1084.

Houston, B. P., Riley, P. A. and Wolfendale, A. W. (1983). *Proc. 18th Int. Cosmic Ray Conf.*, Bangalore, **1**, 89.

Houston, B. P. and Wolfendale, A. W. (1982). *Vistas Astr.*, **26**, 107.

Houston, B. P. and Wolfendale, A. W. (1983). *Astron. Astrophys.*, **126**, 22.

Houston, B. P. and Wolfendale, A. W. (1984*a*). *J. Phys. G.*, **10**, 1587.

Houston, B. P. and Wolfendale, A. W. (1985). *J. Phys. G.*, **11**, 407.

Houston, B. P., Wolfendale, A. W. and Young, E. C. M. (1984). *J. Phys. G.*, **10**, L147.

Hutchinson, G. W. (1952). *Phil. Mag.*, **43**, 847.

Issa, M. R., Strong, A. W. and Wolfendale, A. W. (1981). *J. Phys. G.*, **7**, 565.

Issa, M. R., Riley, P. A., Strong, A. W. and Wolfendale, A. W. (1981). *J. Phys. G.*, **7**, 973.

Issa, M. R. and Wolfendale, A. W. (1981*a*). *J. Phys. G.*, **7**, L187.

Issa, M. R. and Wolfendale, A. W. (1981*b*). *Nature*, **292**, 430.

Khavtassi, J. Sh. (1960). *Atlas of Galactic Dark Nebulae*, Tiblisi, Abastumani Astrophysics Observatory.

Kniffen, D. A. and Fichtel, C. E. (1970). *Astrophys. J. Lett.*, **161**, L157.

Kniffen, D. A., Hartman, R. C., Thompson, D. J., Bignami, G. F., Fichtel, C. E., Ogelman, H. and Tumer, T. (1974). *Nature*, **251**, 397.

Kniffen, D. A., Bertsch, D. L., Morris, D. L., Palmeira, R. A. R. and Rao, K. R. (1978). *Astrophys. J.*, **225**, 591.

Kniffen, D. A., Fichtel, C. E. and Hartman, R. C. (1983). *Proc. 18th Int. Conf. Cosmic Ray*, Bangalore, **1**, 165.

Knight, F. K. (1981). Ph.D. Thesis, University of San Diego.

Kraushaar, W. L., Clark, G. W., Garmire, G. P., Helmken, H., Higbie, P. and Agogino, M. (1965). *Astrophys. J.*, **141**, 845.

Kraushaar, W. L., Clark, G. W., Garmire, G. P., Borken, R., Higbie, P., Leong, C. and Thoros, J. (1972). *Astrophys. J.*, **177**, 341.

Kutner, M. L., Tucker, K. D., Chin, G. and Thaddeus, P. (1977). *Astrophys. J.*, **215**, 521.

Lamb, R. C., Fichtel, C. E., Hartman, R. C., Kniffen, D. A. and Thompson, D. J. (1977). *Astrophys. J. Lett.*, **212**, L63.

Lebrun, F. and Paul, J. A. (1979). *Proc. 16th Int Cosmic Ray Conf.*, Kyoto, **12**, 13.

Lebrun, F. and Paul, J. A. (1983), *Astrophys. J.*, **266**, 276.

Lebrun, F., Bignami, G. F., Buccheri, R., Caraveo, P. A., Hermsen, W., Kanbach, G., Mayer-Hasselwander, H. A., Paul, J. A., Strong, A. W. and Wills, R. D. (1982). *Astron. Astrophys.*, **107**, 390.

Lebrun, F. and Paul, J. A. (1983). *Astrophys. J.*, **266**, 276.

Li, T. P. and Wolfendale, A. W. (1981). *Astron. Astrophys. Lett.*, **100**, L26.

Li, T. P., Riley, P. A. and Wolfendale, A. W. (1983). *Mon. Not. Roy. Astr. Soc.*, **203**, 87.

Lichti, G. G., Buccheri, R., Caraveo, P., Gerardi, G., Hermsen, W., Kanbach, G., Masnou, J. L., Mayer-Hasselwander, H. A., Paul, J. A., Swanenburg, B. N. and Wills, R. D. (1980). *Advances in Space Exploration*, Vol. 7, p. 49, Pergamon Press, Oxford.

Masnou, J. L. (1977). Proc. ESLAB Symp., *Recent Advances in Gamma Ray Astronomy*, p. 33, ESA SP-124.

Mayer-Hasselwander, H. A., Bennett, K., Bignami, G. F., Buccheri, R., Caraveo, P. A., Hermsen, W., Kanbach, G., Lebrun, F., Lichti, G. G., Masnou, J. L., Paul, J. A., Pinkau, K., Sacco, B., Scarsi, L., Swanenburg, B. N. and Wills, R. D. (1982). *Astron. Astrophys.*, **103**, 164.

Montmerle, Th. (1979). *Astrophys. J.*, **231**, 95.

Montmerle, Th. (1982). *Phil. Trans. R. Soc. Lond.*, **A301**, 505.

Morfill, G. E., Volk, H. J., Drury, L., Forman, M., Bignami, G. F. and Caraveo, P. A. (1981). *Astrophys. J.*, **246**, 810.

Ogelman, H. B., Fichtel, C. E., Kniffen, D. A. and Thompson, D. J. (1976). *Astrophys. J.*, **209**, 584.

Omnes, R. (1969). *Phys. Rev. Lett.*, **23**, 38.

Pagel, B. E. J. and Edmunds, M. G. (1981). *Ann. Rev. Astron. Astrophys. J.*, **276**, 182.

Parsignault, D. R., Gursky, H., Kellogg, E. M., Matilsky, T., Murray, S., Schreir, E., Tananbaum, H. and Giacconi, R. (1972). *Nature Phys. Sci.*, **239**, 123.

Parsignault, D. R., Schreier, E., Grindlay, J. and Gursky, H. (1976). *Ap. J. Lett.*, **209**, L73.

Pollock, A. M. T., Bignami, G. F., Hermsen, W., Kanbach, G., Lichti, G. G., Masnou, J. L., Swannenburg, B. N. and Wills, R. D. (1981). *Astron. Astrophys.*, **94**, 116.

Poon, C. B. (1983). Ph.D. Thesis, University of Hong Kong.

Protheroe, R. J., Strong, A. W. and Wolfendale, A. W. (1979). *Mon. Not. Roy. Astr. Soc.*, **188**, 863.

Rana, N. C., Sadzinska, M., Wdowczyk, J. and Wolfendale, A. W. (1984). *Astron. Astrophys.*, **141**, 394.

Riley, P. A., Wolfendale, A. W., Xu, C. X., Manchester, R. N., Robinson, B. J. and Whiteoak, J. B. (1984). *Mon. Not. Roy. Astr. Soc.*, **206**, 423.

Sacher, W. and Schonfelder, V. (1984). *Astrophys. J.*, **279**, 817.

Said, S. S., Wolfendale, A. W., Giler, M. and Wdowczyk, J. (1982). *J. Phys. G.*, **8**, 383.

Sanders, D. B., Solomon, P. M. and Scoville, N. Z. (1984). *Astrophys. J.*, **276**, 182.

Scarsi, L., Buccheri, R., Gerardi, G. and Sacco, B. (1981). *Origin of Cosmic Rays*, IAU Symp. No. 94, p. 279 (Setti, G., Spada, G. and Wolfendale, A. W. eds), Reidel, Dordrecht.

Schlickeiser, R. (1981). *Astron. Astrophys.*, **94**, 57.

Scoville, N. Z. and Solomon, P. M. (1973). *Astrophys. J.*, **180**, 31.

Scoville, N. Z., Solomon, P. M. and Jefferts, K. (1974). *Astrophys. J. Lett.*, **187**, L63.

Share, G. H. (1973). *Int. Symp. and Workshop on Gamma-Ray Astrophysics*, Goddard Space Flight Centre, X-641-73-180.

Share, G. H., Kinzer, R. L. and Seeman, N. (1973). *X-rays and Gamma Ray Astronomy*, Proc. IAU Symp. 55 (Madrid) (Bradt, H. and Giacconi, R. eds), Reidel, Dordrecht.

Solomon, P. M. and Stecker, F. W. (1974). *Proc. ESLAB Gamma Ray Symp.*, Frascati, ESRO SP-106, 253.

Stecker, F. W. (1969). *Astrophys. J.*, **157**, 507.

Stecker, F. W. (1969). *Nature*, **22**, 865.

Stecker, F. W. (1971), *Cosmic Gamma Rays*, Mono Book Corp., Baltimore.

Stecker, F. W. (1971). *Cosmic Gamma Rays*, NASA SP-249, Washington, DC, U.S. Govt. Print. Off.

Stecker, F. W. (1978). *Nature, Lond.*, **273**, 493.

Strong, A. W., Wdowczyk, J. and Wolfendale, A. W. (1973). *Gamma Ray Astrophysics*, p. 259, NASA SP-339.

Strong, A. W., Wdowczyk, J. and Wolfendale, A. W. (1974). *J. Phys. A*, **7**, 120.

Strong, A. W., Wolfendale, A. W. and Worral, D. M. (1976). *J. Phys. A*, **9**, 1553.

Strong, A. W. and Wolfendale, A. W. (1981). *Phil. Trans. Roy. Soc. Lond.*, **A301**, 541.

Strong, A. W., Bignami, G. F., Bloemen, J. B. G. M., Buccheri, R., Caraveo, P. A., Hermsen, W., Kanbach, G., Lebrun, F., Mayer-Hasselwander, H. A., Paul, J. A. and Wills, R. D. (1982). *Astron. Astrophys.*, **115**, 404.

Strong, A. W. and Lebrun, F. (1982). *Astron. Astrophys.*, **105**, 159.

Strong, A. W. and Bignami, G. F. (1983). *Astrophys. J.*, **274**, 549.

Swannenburg, B. N., Bennett, K., Bignami, G. F., Buccheri, R., Caraveo, P., Hermsen,

W., Kanbach, G., Lichti, G. G., Masnou, J. L., Mayer-Hasselwander, H. A., Paul, J. A., Sacco, B., Scarsi, L. and Wills, R. D. (1981). *Astrophys. J. Lett.*, **243**, L69.

Thompson, D. J., Fichtel, C. E., Kniffen, D. A. and Ogelman, H. B. (1975). *Astrophys. J. Lett.*, **200**, L79.

Thompson, D. J., Fichtel, C. E., Hartman, R. C., Kniffen, D. A., Bignami, G. F., Lamb, R. C., Ogelman, H., Ozel, M. E. and Tumer, T. (1976). *The Structure and Content of the Galaxy and Galactic Gamma Rays*, p. 1, Goddard Space Flight Centre X-662-76-154.

Thompson, D. J., Fichtel, C. E., Kniffen, D. A. and Ogelman, H. B. (1977). *Astrophys. J. Lett.*, **214**, L17.

Thompson, D. J. and Fichtel, C. E. (1982). *Astron. Astrophys.*, **109**, 352.

Trombka, J. I., Dyker, C. S., Evans, L. G., Bielefield, H. M., Seltzer, S. M. and Metzger, A. E. (1977). *Astrophys. J.*, **212**, 925.

Trombka, J. I. and Fichtel, C. E. (1982). NASA Tech. Memo. 84928, Goddard Space Flight Centre, Maryland.

Wills, R. D., Bennett, K., Bignami, G. F., Buccheri, R., Caraveo, P., D'Amino, N., Hermsen, W., Kanbach, G., Lichti, G. G., Masnou, J. L., Mayer-Hasselwander, H. A., Paul, J. A., Sacco, B. and Swanenburg, B. N. (1980). *Non-solar gamma-rays* (Cospar Symp.) (Cowsik, R. and Wills, R. D. eds). *Adv. Space Explor.*, **7**, 43.

Wills, R. D. (1981). *Phil. Trans. Roy. Soc. Lond.*, **A301**, 537.

Wills, R. D., Bennett, K., Bignami, G. F., Buccheri, R., Caraveo, P. A., Hermsen, W., Kanbach, G., Masnou, J. L., Mayer-Hasselwander, H. A., Paul, J. A. and Sacco, B. (1982). *Nature*, **296**, 723.

Wolfendale, A. W. (1980). *IAU Symp. 94*, p. 309, D. Reidel, Dordrecht.

Worral, D. M. (1977). Ph.D. Thesis, University of Durham.

Wouterloot, J. G. A. (1981). Ph.D. Thesis, Leiden University.

Chapter 5

Bartelt, J., Courant, H., Heller, K., Joyce, T., Ayres, D. S., Dawson, J., Fields, T., May, E. N., Price, L. E. and Sivaprasad, K. (1985). Preprint DOE/ER/40105-506, ANL-HEP-PR-84-8C.

Bhat, P. N., Gupta, S. K., Ramana Murthy, P. V., Streekantan, B. V., Tonwar, S. C. and Vishwanath, P. R. (1980). *Astron. Astrophys. Lett.*, **81**, L3.

Bhat, P. N., Gopalkrishnan, N. V., Gupta, S. K., Ramana Murthy, P. V., Sreekantan, B. V., Tonwar, S. C. and Vishwanath, P. R.(1982). *Mon. Not. Roy. Astron. Soc.*, **199**, 1007.

Bhat, C. L., Sapru, M. L. and Kant, C. L. (1980). *Nature*, **288**, 146 (*corrigendum*, 1981, 291, 168).

Bhat, C. L., Sapru, M. L. and Razdan, H. (1985). Preprint.

Bignami, G. F., Caraveo, P. and Paul, J. A. (1984). *Nature*, **310**, 464.

Blackett, P. M. S. (1948). Report Gassiot Committee 34, UK.

Boone, J., Cady, R., Cassiday, G. L., Elbert, J. W., Loh, E. C., Sokolsky, P., Steck, D. and Wasserbaech, S. (1984). *Astrophys. J.*, **285**, 264.

Bowyer, C. S., Lampton, M., Mack, J. and De Mondonca, F. (1970). *Astrophys. J. Lett.*, **161**, L1.

Braun, O. and Sitte, K. (1965). *Proc. 9th Int. Cosmic Ray Conf.*, 2, 712.

Browning, R. and Turver, K. E. (1977). *Nuovo Cimento*, **38A**, 223.

Caraveo, P., Bignami, G. F., Giommi, P., Merghetti, S. and Paul, J. A. (1984). *Nature*, **310**, 481.

Carr, B. J. (1976). *Astrophys. J.*, **206**, 8.

Cawley, M. F., Clear, J., Fegan, D. J., Gibbs, K., Porter, N. A. and Weekes, T. C. (1982). *Proc. Intern. W/S on Very High Energy Gamma Ray Astronomy*, Ootacamund, p. 292 (Ramana Murthy, P. V. and Weekes, T. C. eds).

Cawley, M. F., Clear, J., Fegan, D. J., Gibbs, K., Gorham, P., Lamb, R. C., Macrae, I., Mackeown, P. K., Porter, N. A., Stenger, V. J., Turver, K. E. and Weekes, T. C. (1983). *Proc. 18th Int. Cosmic Ray Conf.*, Bangalore, 9, 61.

Cawley, M. F. and Weekes, T. C. (1984). Preprint 1943, Centre for Astrophysics, Cambridge, USA.

Chantler, M. P., Dowthwalte, J. C., Harrison, A. B., Kirkman, I. W., Macrae, H. J., Orford, K. J., Turver, K. E. and Walmsley, M. S. (1984). Preprint.

Chatterjee, B. K., Murthy, G. T., Ramana Murthy, P. V., Sreekantan, B. V. and Tonwar, S. C. (1971). *Nature Phys. Sci.*, **231**, 126.

Cheng, A., Ruderman, M. and Sutherland, P. (1976). *Astrophys. J.*, **203**, 209.

Chudakov, A. E., Zatsepin, V. I., Nesterova, N. M. and Dadykin, V. L.. (1962). *J. Phys. Soc. Jpn.*, **17**, (A-III), 106.

Chudakov, A. E., Dadykin, V. L., Zatsepin, V. I. and Nesterova, N. M. (1964). *Proc. Lebedev Physical Inst.*, **26**, 118.

Chudakov, A. E., Dadykin, V. L., Zatsepin, V. I. and Nesterova, N. M. (1965). *Trans. Consultants Bureau*, **26**, 99.

Clay, R. W., Protheroe, R. J. and Gerhardy, P. R. (1984). *Nature*, **309**, 687.

Cocconi, G. (1959). *Proc. Moscow Conf. Cosmic Rays*, **2**, 309.

Cocke, W. J., Disney, M. J. and Taylor, D. J. (1969). *Nature*, **221**, 525.

Comella, J. M., Craft, H. D., Lovelace, R. V. E., Sutton, J. M. and Tyler, G. L. (1969). *Nature*, **221**, 453.

Craig, M. A. B., Orford, K. J., Turver, K. E. and Weekes, T. C. (1981). *Proc. 17th Int. Cosmic Ray Conf.*, Paris, **1**, 3.

Danaher, S., Fegan, D. J., Porter, N. A. and Weekes, T. C. (1981). *Nature*, **289**, 568.

Dowthwaite, J. C., Gibson, A. I., Harrison, A. B., Kirkman, I. W., Lotts, A. P., Macrae, J. H., Orford, K. J., Turver, K. E. and Walmsley, M. (1983). *Astron. Astrophys.*, **126**, 1.

Dowthwaite, J. C., Harrison, A. B., Kirkman, I. W., Macrae, J. H., Orford, K. J., Turver, K. E. and Walmsley, M. S. (1984*a*). *Astron. Astrophys. Lett.*, **136**, L14.

Dowthwaite, J. C., Harrison, A. B., Kirkman, I. W., Macrae, J. H., Orford, K. J., Turver, K. E. and Walmsley, M. S. (1984*b*). *Astron. Astrophys.*, **142**, 55.

Dowthwaite, J. C., Harrison, A. B., Kirkman, I. W., Macrae, J. H., Orford, K. J., Turver, K. E. and Walmsley, M. S. (1984*c*). *Astrophys. J. Lett.*, **286**, L35.

Dowthwaite, J. C., Harrison, A. B., Kirkman, I. W., Macrae, J. H., Orford, K. J., Turver, K. E. and Walmsley, M. S. (1984*d*). *Nature*, **309**, 691.

Dzikowski, T., Gawin, J., Grochalska, B., Wasilewski, A. and Wdowczyk, J. (1981). *Proc. 17th Int. Cosmic Ray Conf.*, Paris, **1**, 8.

Dzikowski, T., Gawin, J., Grochalska, B. and Wdowczyk, J. (1983). *J. Phys. G.*, **9**, 459.

Eichler, D. and Vestrand, W. T. (1984). *Nature*, **307**, 613.

Erickson, R. A., Fickle, R. K. and Lamb, R. C. (1976). *Astrophys. J.*, **210**, 539.

Fazio, G. G. (1967). *Ann. Rev. Astron. & Astrophys.*, **5**, 481.

Fazio, G. G., Helmken, H. F., O'Mongain, E., Rieke, G. H. and Weekes, T. C. (1972). *Astrophys. J. Lett.*, **175**, L117.

Felten, J. E. and Morrison, P. (1966). *Astrophys. J.*, **146**, 686.

Fomin, V. P., Vladimirsky, B. M. and Stepanian, A. A. (1977). *Proc. 15th Int. Cosmic Ray Conf.*, Plovdiv, **1**, 12.

Fritz, G., Henry, R. C., Meekins, J. F., Chubb, T. A. and Friedman, H. (1969). *Science*, **164**, 709.

Fruin, J. H., Jelley, J. V., Long, C. D., Porter, N. A. and Weekes, T. C. (1964). *Phys. Lett.*, **10**, 176.

Galbraith, W. (1958). *Extensive Air Showers*, Butterworths, London.

Galbraith, W. and Jelley, J. V. (1953). *Nature*, **171**, 349.

Galbraith, W. and Jelley, J. V. (1955). *J. Atmos. Terr. Phys.*, **6**, 250.

Galper, A. M., Kirillov-Ugryumov, V. G., Kurochkin, A. V., Leikov, P. G., Luchkov, B. I., Urkin, U. L., Fomin, V. P., Neshpor, Y. I., Stepanian, A. A. and Vladimirsky, B. M. (1977). *Proc. 15th Int. Cosmic Ray Conf.*, Plovdiv, **1**, 131.

Gawin, J., Hibner, J., Zawadzki, A. and Maze, R. (1963). *Proc. 8th Int. Cosmic Ray Conf.*, Jaipur, **4**, 180.

Gibson, A. I., Harrison, A. B., Kirkman, I. W., Lotts, A. P., Macrae, J. H., Orford, K. J., Turver, K. E. and Walmsley, M. S. (1982*a*). *Proc. Int. W/S on Very High Energy Gamma Ray Astronomy*, Ootacamund, p. 97 (Ramana Murthy, P. V. and Weekes, T. C. eds).

Gibson, A. I., Harrison, A. B., Kirkman, I. W., Lotts, A. P., Macrae, J. H., Orford, K. J., Turver, K. E. and Walmsley, M. S. (1982*b*). *Nature*, **296**, 833.

Ginzburg, V. L. and Syrovatskii, S. I. (1964). *The Origin of Cosmic Rays*, Pergamon Press, New York.

Goldreich, P. and Julian, W. H. (1969). *Astrophys. J.*, **157**, 869.

Gorham, P. W., Stenger, V. J. and Weekes, T. C. (1982). *Proc. Int. W/S Very High Energy Gamma Rays*, Ootacamund, p. 288 (Ramana Murthy, P. V. and Weekes, T. C. eds).

Gould, R. J. and Burbidge, G. R. (1965). *Ann. d'Astrophys.*, **28**, 171.

Gould, R. J. and Schreder, G. P. (1967). *Phys. Rev.*, **155**, 1408.

Greisen, K. (1960). *Ann. Rev. Nucl. Sci.* **10**, 63.

Grindlay, J. E. (1971). SAO Special Report no. 334, Smithsonian Astrophysical Observatory, USA.

Grindlay, J. E. (1972). *Proc. Int. W/S on Very High Energy Gamma Ray Astronomy*, Ootacamund, p. 178 (Ramana Murthy, P. V. and Weekes, T. C. eds).

Grindlay, J. E., Helmken, H. F., Hanbury Brown, R., Davis, J. and Allen, L. R. (1975*a*). *Astrophys. J. Lett.*, **197**, L9.

Grindlay, J. E., Helmken, H. F., Hanbury Brown, R., Davis, J. and Allen, L. R. (1975*b*). *Astrophys. J.*, **201**, 82.

Grindlay, J. E., Helmken, H. F. and Weekes, T. C. (1976). *Astrophys. J.*, **209**, 592.

Gunn, J. E. and Ostriker, J. P. (1969). *Nature*, **221**, 454.

Gupta, S. K. (1983). Ph.D. Thesis, University of Bombay, India, unpublished.

Gupta, S. K., Ramana Murthy, P. V., Sreekantan, B. V. and Tonwar, S. C. (1978). *Astrophys. J.*, **221**, 268.

Gupta, S. K., Ramana Murthy, P. V., Sreekantan, B. V., Tonwar, S. C. and Viswanath, P. R. (1982). *Proc. Int. W/S on Very High Energy Gamma Ray Astronomy*, Ootacamund, p. 279 (Ramana Murthy, P. V. and Weekes, T. C. eds).

Hayashida, N., Ishikawa, F., Kamta, K., Kifune, T., Nagano, M. and Tan, Y. H. (1981). *Proc. 17th Int. Cosmic Ray Conf.*, Paris, **9**, 9.

Helmken, H. F. and Weekes, T. C. (1979). *Astrophys. J.*, **228**, 531.

Hillas, A. M. (1984). *Nature*, **312**, 50.

Hillier, R. R., Jackson, W. R., Murray, A., Redfern, R. M. and Sale, R. G. (1970). *Astrophys. J.*, **206**, 282.

Hinata, S. (1977). *Astrophys. J.*, **216**, 101.

Jelley, J. V. (1967). In *Prog. Elementary Particles and Cosmic Ray Physics*, vol. **9**, p. 41 (Wilson, J. G. and Wouthuysen, S. A. eds), North Holland Publishing Co., Amsterdam.

Jelley, J. V. and Porter, N. A. (1963). *Quart. J. Roy. Astr. Soc.*, **4**, 275.

Jennings, D. M., White, G., Porter, N. A., O'Mongain, E., Fegan, D. J. and White, J. (1974). *Nuovo Cimento*, **20**, 71.

Kanbach, G., Bennett, K., Bignami, G. F., Buccheri, R., Caraveo, P., D'Amico, N., Hermsen, W., Lichti, G. G., Masnou, J. L., Mayer-Hasselwander, H. A., Paul, J. A., Sacco, B., Swanenburg, B. N. and Wills, R. D. (1980). *Astron. Astrophys.*, **90**, 163.

Lamb, R. C., Fichtel, C. E., Hartman, R. C., Kniffen, D. A. and Thompson, D. J. (1977). *Astrophys. J. Lett.*, **212**, L63.

Lamb, R. C., Godfrey, C. P., Wheaton, W. A. and Turner, T. (1982). *Nature*, **296**, 543.

Lambert, A., Lloyd-Evans, J. and Watson, A. A. (1983). *Proc. 18th Int. Cosmic Ray Conf.*, Bangalore, **2**, 144.

Lichti, G. G., Buccheri, R., Caraveo, P., Gerardi, G., Hermsen, W., Kanbach, G., Masnou, J. L., Mayer-Hasselwander, H. A., Paul, J. A., Swanenburg, B. N. and Wills, R. D. (1980). In *Advances Space Exploration*, vol. 7, p. 49 (Cowsik, R. and Wills, R. D. eds), Pergamon Press, Oxford.

Linsley, J. (1981). In *Origin of Cosmic Rays*, p. 53 (Setti, G., Spada, G. and Wolfendale, A. W. eds), IAU Publication.

Lloyd-Evans, J., Coy, R. N., Lambert, A., Lapikens, J., Patel, M., Reid, R. J. D. and Watson, A. A. (1983*a*). *Nature*, **305**, 784.

Lloyd-Evans, J., Coy, R. N., Lambert, A., Lapikens, J., Patel, M., Reid, R. J. D. and Watson, A. A. (1983*b*). *Proc. 18th Int. Cosmic Ray Conf.*, Bangalore, **9**, 65.

Long, C. D., McBreen, B., Porter, N. A. and Weekes, T. C. (1965). *Proc. 9th Int. Cosmic Ray Conf.*, London, **1**, 318.

Manchester, R. N. and Taylor, J. H. (1977). *Pulsars*, Freeman, San Francisco.

Massaro, E. and Salvati, M. (1979). *Astron. & Astrophys.*, **71**, 51.

Michel, F. C. (1982). *Rev. Mod. Phys.*, **54**, 1.

Milgrom, M. and Pines, D. (1978). *Astrophys. J.*, **220**, 272.

Morello, C., Navarra, G. and Vernetto, S. (1983). *Proc. 18th Int. Cosmic Ray Conf.*, Bangalore, **1**, 127.

Mukanov, D. B., Nesterova, N. M., Stepanian, A. A. and Fomin, V. P. (1980). *Izv. Krimskoi Astrofiz Obs.*, **62**, 98.

Neshpor, Y. I., Stepanian, A. A., Fomin, V. P., Gerasimov, S. A., Vladimirsky, B. M. and Ziskin, Y. L. (1979). *Astrophys. & Space Sci.*, **61**, 349.

Nishimura, J. and Kamata, K. (1952). *Prog. Theor. Phys.*, **7**, 185.

Ostriker, J. P. (1972). In *The Physics of Pulsars*, p. 159 (Lenchek, A. M. ed.), Gordon and Breach, New York.

Page, D. N. and Hawkin, S. W. (1976). *Astrophys. J.*, **206**, 1.

Parsignault, D. R., Schreier, E., Grindlay, J. and Gursky, H. (1976). *Astrophys. J. Lett.*, **209**, L73.

Porter, N. A. (1983). *Proc. 18th Int. Cosmic Ray Conf.*, Bangalore, **12**, 435.

Porter, N. A., Delaney, T., Helmken, H. F. and Weekes, T. C. (1976). *Nuovo Cimento*, **32B**, 514.

Porter, N. A. and Weekes, T. C. (1978). SAO Special Report no. 381, Smithsonian Astrophysical Observatory, USA.

Porter, N. A. and Weekes, T. C. (1979). *Nature*, **277**, 199.

Protheroe, R. J., Clay, R. W. and Gerhardy, P. R. (1984). *Astrophys. J. Lett.*, **280**, L47.

Radhakrishnan, V. and Srinivasan, G. (1981). Unpublished.

Ramana Murthy, P. V. (1980). In *Non-Solar Gamma-Rays (COSPAR)*, p. 71 (Cowsik, R. and Wills, R. D. eds), Pergamon Press, Oxford.

Ramana Murthy, P. V. and Tonwar, S. C. (1976). Optical night sky background at Ootacamund, Tamilnadu, India, unpublished.

Rana, N. C., Sadzinska, M., Wdowczyk, J. and Wolfendale, A. W. (1984). *Astron. Astrophys.*, **141**, 394.

Rieke, G. H. and Weekes, T. C. (1969). *Astrophys. J.*, **155**, 429.

Ruderman, M. A. and Sutherland, P. G. (1975). *Astrophys. J.*, **196**, 51.

Samorski, M. and Stamm, W. (1983*a*). *Astrophys. J. Lett.*, **268**, L17.

Samorski, M. and Stamm, W. (1983*b*). *Proc. 18th Int. Cosmic Ray Conf.*, Bangalore, **1**, 135.

Samorski, M. and Stamm, W. (1983*c*). *Proc. 18th Int. Cosmic Ray Conf.*, Bangalore, **11**, 244.

Shabad, A. E. and Usov, V. V. (1982). *Nature*, **295**, 215.

Shklovskii, I. S. (1953). *Dokl. Acad. Nauk SSR*, **90**, 983.

Smith, F. G. (1977). *Pulsars*, Cambridge University Press, Cambridge.

Sreekantan, B. V. (1981). *Phil. Trans. Roy. Soc. Lond.*, **A301**, 629.

Staelin, D. H. and Reifenstein, E. C. (1968). *Science*, **162**, 1481.

Stamenov, J. N., Ushev, S. Z., Nikolskii, S. I. and Yakovlev, V. I. (1983). *Proc. 18th Int. Cosmic Ray Conf.*, Bangalore, **6**, 54.

Stamm, W. and Samorski, M. (1983). *Proc. 18th Int. Cosmic Ray Conf.*, Bangalore, **1**, 131.

Stecker, F. W. (1971). *Cosmic Gamma Rays*, NASA publication no. SP 249, USA.

Stepanian, A. A. (1982). *Astrophys. & Space Sci.*, **84**, 347.

Stepanian, A. A., Fomin, V. P., Nesphor, Y. I., Vladimirsky, B. M. and Zyskin, Y. L. (1982). *Proc. Int. W/S Very High Energy Gamma Ray Astronomy*, Ootacamund, p. 43 (Ramana Murthy, P. V. and Weekes, T. C. eds).

Stephens, S. A. and Verma, R. P. (1984). *Nature*, **308**, 828.

Stephenson, C. B. and Sanduleak, N. (1977). *Astrophys. J. Suppl.*, **33**, 459.

Sturrock, P. A. (1971). *Astrophys. J.*, **164**, 529.

Swanenburg, B. N., Bennett, K., Bignami, G. F., Buccheri, R., Caraveo, P., Hermsen, W., Kanbach, G., Lichti, G. G., Masnou, J. L., Mayer-Hasselwander, H. A., Paul, J. A., Sacco, B., Scarsi, L. and Wills, R. D. (1981). *Astrophys. J. Lett.*, **243**, L69.

Toyoda, Y., Suga, K., Murakami, K., Hasegawa, H., Shibata, S., Domingo, V., Escobar, I., Kamata, K., Bradt, H., Clark, G. and La Pointe, M. (1965). *Proc. 9th Int. Cosmic Ray Conf.*, London, **2**, 708.

Turver, K. E. and Weekes, T. C. (1978). *Nuovo Cimento*, **458**, 99.

van der Kliss, M. and Bonnet-Bidaud, J. M. (1981). *Astron. Astrophys. Lett.*, **95**, L5.

Vestrand, W. T. and Eichler, D. (1982). *Astrophys. J.*, **261**, 251.

Vishwanath, P. R. (1982). *Proc. Int. W/S on Very High Energy Gamma Ray Astronomy*, Ootacamund, p. 21 (Ramana Murthy, P. V. and Weekes, T. C. eds).

Vishwanath, P. R., Gupta, S. K., Ramana Murthy, P. V., Sreekantan, B. V. and Tonwar, S. C. (1985). Preprint.

Vladimirsky, B. M., Stepanian, A. A. and Fomin, V. P. (1973). *Proc. 13th Int. Cosmic Ray Conf.*, Denver, **1**, 456.

Wdowczyk, J. (1965). *Proc. 9th Int. Cosmic Ray Conf.*, London, **2**, 691.

Wdowczyk, J. and Wolfendale, A. W. (1983). *Nature*, **305**, 609.

Wdowczyk, J. and Wolfendale, A. W. (1984). *J. Phys. G*, **10**, 1599.

Weekes, T. C., Fazio, G. G., Helmken, H. F., O'Mongain, E. P. and Rieke, G. H. (1972). *Astrophys. J.*, **174**, 165.

Weekes, T. C. and Helmken, H. F. (1977). *Recent Adv. Gamma Ray Astron. Conf.*, ESA (Frascati), 39.

Weekes, T. C., Helmken, H. F. and Grindlay, J. E. (1979). *Proc. 16th Int. Cosmic Ray Conf.*, Kyoto, **1**, 132.

Zyskin, L. and Mukanov, D. B. (1983). *Proc. 18th Int. Cosmic Ray Conf.*, Bangalore, **1**, 122.

INDEX